"This is an excellent and timely review of the social psychology and forensic practice. The w ging, the examples vivid, and the implications gone beyond the boundaries that most books on this topic cover, making it truly unique, generative, and worth having on your bookshelf."

Kipling D. Williams, Distinguished Professor, Department of Psychological Sciences, Purdue University, USA

"This gem of a text invites us to shine a light through the many layers of complexity in forensic practice, providing exciting new social psychological perspectives on critical issues: power, attribution, ostracism, impression management, to name a few. From time to time, the unique world of forensic services feels like a turbulent, risk-saturated tinderbox, an imbalanced and imperfect world where the majority are typically related to as less powerful than the minority who hold the keys. This volume provides a much-needed demonstration, brought to life with case material throughout, of how to extend the relational, more culturally competent tool-kit for forensic practitioners; to shift shared understandings and consequent dialogues more robustly in the direction of change where it is needed most."

Estelle Moore, Associate Professor of Forensic Psychology, Clinical & Forensic Psychologist, Head of High Secure Psychological Services, Broadmoor Hospital, UK

Social Psychology in Forensic Practice

This book explores how different social psychology theories and concepts can be applied to practice. Considering theories from attribution theory to coercion theory, social identity theories to ostracism, the authors offer a greater understanding and appreciation of the ways in which social psychology can contribute to forensic practice.

The book argues that social psychology is useful for carrying out assessments (including risk assessments), formulations, and interventions with clients in forensic settings, as well as for psychological consultation, training, and the development of services. These theories are also important when understanding multi-disciplinary and multi-agency working, staff–client relationships, and peer-to-peer relationships. Through illustrative composite case examples, taken from the authors' experiences in forensic settings, the chapters demonstrate effective ways to pursue a theoretically informed practice.

Exploring a broad range of theories and a timely topic, *Social Psychology in Forensic Practice* will interest a wide readership including graduate and undergraduate students and researchers in criminology, sociology, and forensic, social and clinical psychology. It will also be of practical use to health professionals and non-health professionals working in forensic settings as well as policy makers and others commissioning forensic services.

Joel Harvey is a senior lecturer in forensic psychology at the Department of Law and Criminology, Royal Holloway, University of London, UK. He holds a PhD in criminology from the Institute of Criminology, University of Cambridge, and a doctorate in clinical psychology from the University of Manchester. He has previously published *Young Men in Prison: Surviving and Adapting to Life Inside* (2007/2012), and co-edited *Psychological Therapy in Prisons and Other Secure Settings* (2010) and *Young People in Forensic Mental Health Settings: Psychological Thinking and Practice* (2015).

Derval Ambrose is a consultant forensic psychologist and the lead psychologist for the Offender Personality Disorder service in South London and Maudsley NHS Foundation Trust. She has worked and held leaderships posts across several organisations, including the National Health Service, HM Prison Service, London Probation Service, the National Offender Management Service and the third sector. Additionally, she is an honorary visiting associate at the Institute of Psychiatry, Psychology and Neuroscience at King's College London, UK.

Social Psychology in Forensic Practice

Edited by
Joel Harvey and Derval Ambrose

Routledge
Taylor & Francis Group
LONDON AND NEW YORK

Cover image: "Rainbow – A Community in Technicolor" Photograph taken by Josip Lizatovic, Creative Arts Technician, June 2020

Displayed in the Healing Garden, River house, Bethlem Royal Hospital. A rainbow collage created by staff and service users in the Forensic Service, to give thanks for all the efforts made during the Covid 19 pandemic. The rainbow remained on display throughout the pandemic and achieved acclaim in the South London and Maudsley NHS "Listening in Action" competition.

First published 2023
by Routledge
4 Park Square, Milton Park, Abingdon, Oxon OX14 4RN

and by Routledge
605 Third Avenue, New York, NY 10158

Routledge is an imprint of the Taylor & Francis Group, an informa business

© 2023 selection and editorial matter, Joel Harvey and Derval Ambrose; individual chapters, the contributors

The right of Joel Harvey and Derval Ambrose to be identified as the authors of the editorial material, and of the authors for their individual chapters, has been asserted in accordance with sections 77 and 78 of the Copyright, Designs and Patents Act 1988.

All rights reserved. No part of this book may be reprinted or reproduced or utilised in any form or by any electronic, mechanical, or other means, now known or hereafter invented, including photocopying and recording, or in any information storage or retrieval system, without permission in writing from the publishers.

Trademark notice: Product or corporate names may be trademarks or registered trademarks, and are used only for identification and explanation without intent to infringe.

British Library Cataloguing-in-Publication Data
A catalogue record for this book is available from the British Library

Library of Congress Cataloging-in-Publication Data
Names: Harvey, Joel, editor. | Ambrose, Derval, editor.
Title: Social psychology in forensic practice / edited by Joel Harvey and Derval Ambrose.
Description: Abingdon, Oxon ; New York, NY : Routledge, 2023. | Includes bibliographical references and index.
Identifiers: LCCN 2022028708 (print) | LCCN 2022028709 (ebook) | ISBN 9781138676145 (pbk) | ISBN 9781138676138 (hbk) | ISBN 9781315560243 (ebk)
Subjects: LCSH: Social psychology. | Forensic psychology.
Classification: LCC HM1033 .S647 2023 (print) | LCC HM1033 (ebook) | DDC 302--dc23/eng/20220818
LC record available at https://lccn.loc.gov/2022028708
LC ebook record available at https://lccn.loc.gov/2022028709

ISBN: 978-1-138-67613-8 (hbk)
ISBN: 978-1-138-67614-5 (pbk)
ISBN: 978-1-315-56024-3 (ebk)

DOI: 10.4324/9781315560243

Typeset in Goudy
by Taylor & Francis Books

To Mags and Dan for being their supportive selves; they have been extremely patient with us while we completed this book

Contents

List of illustrations   xi
Foreword   xiii
List of contributors   xv
Acknowledgements   xviii
List of Abbreviations   xix

1 Introduction   1
   JOEL HARVEY AND DERVAL AMBROSE

2 Attributions and Biases   21
   LAURA BOWDEN, EMILY GLORNEY AND EMILY DURBER

3 Social Identity Theories   47
   DEBORAH MORRIS AND ELANOR WEBB

4 Impression Management   69
   JOEL HARVEY AND DEBORAH H. DRAKE

5 Attitudes and Beliefs   91
   LARA ARSUFFI

6 Aggression   116
   MATT BRUCE AND VERONICA ROSENBERGER

7 Group Formation and Behaviour   144
   DERVAL AMBROSE AND TANIA TANCRED

8 Coercion and Social Influence   165
   VYV HUDDY AND TIMOTHY A. CAREY

9 Ostracism   183
   DENNIS KAIP AND JOEL HARVEY

10 Stereotyping and Prejudice   204
    DERVAL AMBROSE, COLIN CAMPBELL AND DENNIS KAIP

    Index   241

Illustrations

Figures

2.1	Multidisciplinary team meeting: examples of possible biases	25
2.2	De-biasing techniques	37
2.3	An illustration of how our 'lenses' may filter our view of the same topic and how our experiences shape our biases	38
5.1	Stages of entry to and exit from the CJS with corresponding potential to shame and reduce dissonance	106
5.2	Opportunities to create dissonance and promote reintegrative shaming in the CJS	106
6.1	The DAMSA for imminent risk prediction and formulation	132
6.2	The DAMSA for dynamic management planning	136
8.1	The basic feedback loop	170
8.2	Interactions between closed casual loops	172
9.1	Multi-layered ostracism	188
10.1	Phinney's stages of ethnic identity development	209
10.2	A therapy dialogue	213

Tables

4.1	Self-states	83
4.2	Discrepancies between self-states and self-guides leading to emotional reactions	83
6.1	Key features of various reactive and proactive theories of aggression	121
6.2	DASA dynamic behaviour items and examples	126
6.3	Psychosocial intervention strategies by aggressor motivation typology	129
6.4	Exploratory questions for eliciting likely motives for aggressor motivations	133
7.1	Symptoms of groupthink	156

Boxes

4.1	Taking an impression management lens in practice	87
5.1	Top tips for effective therapeutic environments	98
5.2	Top tips for changing attitudes and behaviour using learning from dissonance research	104
6.1	Aggression dichotomies	118
6.2	Nursing note regarding Maurice	119
6.3	Aggressive incident	134
6.4	Maurice's reflections of the incident	136
6.5	Serena's experience	137
10.1	Practitioner tip: Phinney's model	211
10.2	Tips for practitioners	221
10.3	Defining gender-responsiveness	223
10.4	Enhancing competence in the context of poverty	229

Foreword

As the editors remind us in their excellent introduction to this volume, social psychology has a long history of examining topics that are of relevance to forensic practice. The most obvious example of this is Zimbardo's 1971 Stanford prison experiment, which was revisited 40 years later by British social psychologists Steve Reicher and Alex Haslam in their BBC prison study. Efforts have been made to celebrate the impact of psychology's role in responding to criminal justice issues – see, for example, the excellent collection of essays on the application of psychology to crime and policing matters, led by social psychologist Clifford Stott (Stott, Bradford, Radburn & Savigar-Shaw, 2020) – but the chapters that follow in this book represent a ground-breaking collective effort to recognise the unique and valuable ways in which social psychology can and does contribute specifically to forensic practice.

Having completed one of the last remaining undergraduate degrees in social psychology (at the University of Sussex) in the late 1990s, it was obvious to me when I later set out to undertake my doctoral studies in forensic psychology that I should draw on a social psychological theoretical framework (for example, Meek, 2007), something I have continued to find valuable in my subsequent research career as a prison scholar. I therefore warmly welcome this book and am glad that the editors have taken the opportunity to showcase the application of social psychological theories and concepts to forensic practice across these ten well-crafted and insightful chapters.

I hope that the impact of this book will not only be a greater understanding and appreciation of the ways in which social psychology can contribute to forensic practice, but also that it will inspire and motivate others to follow suit. Researchers, practitioners, and policy makers should be empowered to seek theoretically driven solutions to criminal justice problems and feel comfortable working within the (sometimes messy) intersection of different disciplines and sub-disciplines in responding more effectively to the challenges we face in society. Research funders are already making efforts to reward such approaches and although psychology is a notoriously divided discipline – reflected in how we teach it, and even in the identities we adopt as psychologists – this exciting collection demonstrates some of the ways in which forensic psychology can draw more usefully from its social psychological neighbour.

Professor Rosie Meek, CPsychol AFBPsS
Department of Law and Criminology
Royal Holloway University of London

References

Meek, R. (2007). The parenting possible selves of young men in prison. *Psychology, Crime and Law*, 13(4), 371–382.

Stott, C., Bradford, B., Radburn, M. & Savigar-Shaw, L. (eds). (2020). *Making an impact on policing and crime: Psychological research, policy and practice*. London: Routledge.

Contributors

Derval Ambrose is a consultant forensic psychologist and the lead psychologist for the Offender Personality Disorder service in South London and Maudsley NHS Foundation Trust. She has worked and held leaderships posts across several organisations, including the National Health Service, HM Prison Service, London Probation Service, the National Offender Management Service and the third sector. Additionally, she is an honorary visiting associate at the Institute of Psychiatry, Psychology & Neuroscience at King's College London. Derval has contributed to three books in the area of forensic psychology and published several articles.

Lara Arsuffi is a forensic psychologist who has worked for the NHS in the UK for the last 20 years. Lara is currently working with high-risk adults on probation orders, who have committed violent or sexual offences, and who are living in the community. Lara is also the module leader for the undergraduate forensic psychology module at Buckingham University, UK. Lara has previously worked with adult men detained in inpatient forensic mental health settings; with high risk teenagers living in the community and managed by the Youth Offending Service; and with children and their parents known to social services due to child protection concerns.

Laura Bowden is a forensic psychologist who has worked in prisons, high, medium and low secure psychiatric hospitals. She currently works as a lecturer in forensic psychology, is a specialist member of the Parole Board, and works in a range of clinical and supervisory roles.

Matt Bruce is an expert witness and assistant professor of forensic psychology at the George Washington University, District of Columbia, USA. He has published numerous peer-reviewed journal articles, empirical studies, and book chapters on the subject of sexual and violence risk assessment, personality disorder, as well as offender treatment and management in forensic setting.

Colin Campbell is associate medical director and consultant forensic psychiatrist, South London and Maudsley NHS Foundation Trust, UK. His clinical interest is in personality disorder and offending behaviour, and he works in Offender Personality Disorder (OPD) pathway services in secure hospital, prison and community settings in south London.

Timothy A. Carey is professor and chair of Country Health Research and Innovation at Curtin University in Western Australia. He is a clinical psychologist and researcher with over 175 publications. He has worked as a professor at universities in Australia, Rwanda, and the US and as a clinical psychologist in the UK and Australia.

Deborah H. Drake is senior lecturer in criminology at the Open University. She has an established research and publication record on prisons and punishment that is internationally recognised. She is author of *Prisons, Punishment and the Pursuit of Security* (Palgrave, 2012) and co-editor of *Palgrave Handbook of Prison Ethnography* (2015).

Emily Durber completed her BSc criminology and psychology at Royal Holloway, University of London. She is currently a postgraduate student studying for a Master's degree in Applied Clinical Psychology at the University of Bath, UK. She works as a Clinical and Therapy Assistant for Novalis Trust, and as a research assistant on a project exploring the use of digital technology for care experienced young people.

Emily Glorney is senior lecturer in forensic psychology at Royal Holloway, University of London, UK. Emily has over 25 years of experience in forensic mental health research and practice, having worked as a practitioner psychologist in a secure hospital for 10 years and since conducting psychological risk assessments in prison.

Joel Harvey is a senior lecturer in forensic psychology at the Department of Law and Criminology, Royal Holloway, University of London. He holds a PhD in criminology from the Institute of Criminology, University of Cambridge, and a doctorate in clinical psychology from the University of Manchester. He has worked in community and secure forensic services. He has previously published *Young Men in Prison: Surviving and Adapting to Life Inside* (2007/2012), and co-edited *Psychological Therapy in Prisons and Other Secure Settings* (2010) and *Young People in Forensic Mental Health Settings: Psychological Thinking and Practice* (2015).

Vyv Huddy is a lecturer in clinical psychology at the University of Sheffield. He has worked in both criminal justice and forensic mental health settings as a clinician and, more recently, has focused on research. His work aims to develop acceptable and feasible psychological therapies and resources for these challenging environments.

Dennis Kaip, BSc (Hons), MSc, has extensive international work experience spanning the last 20 years. Until recently Dennis has been working as a higher assistant clinical psychologist in an English prison and as an honorary research assistant at the Institute of Psychiatry, Psychology and Neuroscience, King's College London. He has been a guest speaker at various Universities sharing his experience working with young people and families as well as unaccompanied asylum seekers and victims of human trafficking in several settings and professional contexts.

Deborah J. Morris, is consultant clinical psychologist and director for the Centre of Developmental and Complex Trauma. Deborah has worked in secure forensic services for over ten years, in clinically and in research capacities. Deborah has published in the areas of risk assessment, personality disorder, trauma, intimate partner violence, occupational trauma, co-production, intellectual disabilities, and moral injury.

Veronica Rosenberger is a Master's-level clinician working in both inpatient and outpatient mental healthcare settings in Central Pennsylvania in the United States. Veronica is passionate about confronting systemic issues in mental healthcare, providing better access to valuable and necessary resources, and handling every client with an individualised and trauma-informed approach.

Tania Tancred is a UK based consultant forensic psychologist, who has worked extensively in the criminal justice system, both in prisons and probation. She specialises in working with high risk, complex men with convictions for violent and sexual crimes, and is currently the Kent lead for the Offender Personality Disorder programme.

Elanor L. Webb is a senior research assistant psychologist and PhD candidate at the Centre for Developmental Complex Trauma, St Andrew's Healthcare, UK. Since completing her studies in clinical psychology research, Elanor has worked in research in a secure psychiatric setting. The primary focus of her work is psychological trauma in staff and service users in such environments.

Acknowledgements

We would like to thank clients and colleagues whom we have worked with over the years who continue to help us learn more about forensic practice. We'd like to thank Emily Durber for her hard work in helping us to copyedit this book and the team at Routledge for seeing it through the press. Thanks too to Adam Danquah and Gerard Drennan for their insightful comments on particular chapters.

Joel would also like to thank Robert Jago, Lisa Smith, Frances MacLennan, and Luke Endersby, for offering support, and Isabel for her insight and expertise in practice, which continues to inspire and shape his thinking. He also thanks his family for their continued support. Derval would like to thank Michelle, Lee, and Zac for their encouragement and valued social distractions during this project. She would also like to thank her parents and family for their continued support.

List of Abbreviations

ACEs	Adverse Childhood Experiences
ACF	Assessment and Classification of Function
ART	Aggression Replacement Training
BAME	Black, Asian, and Minority Ethnic
BME	Black and Minority Ethnic
BVC	Broset Violence Checklist
CBT	Cognitive Behavioural Therapy
CJS	Criminal Justice System
COSA	Circles of Support and Accountability
CPT	Compassion Focused Therapy
CSAP	Correctional Services Accreditation Panel
CSI	Criminal Social Identity
DAMSA	Dynamic Assessment and Management of Situational Aggression
DASA	Dynamic Appraisal of Situational Risk
DRT	Differential Reinforcement Theory
EA	Explicit Attitudes
EMS	Early Maladaptive Schema
GAM	General Aggression Model
GRSP	Gender Role Strain Paradigm
HCR-20	Historical-Clinical-Risk 20
HMPPS	His Majesty's Prison and Probation Service
IA	Implicit Attitudes
IAT	Implicit Associations Test
IPM-CSI	Integrated Psychosocial Model of Criminal Social Identity
IPV	Intimate Partner Violence
LSI-R	Level of Supervision Inventory-Revised
MAPPA	Multi-Agency Public Protection Arrangements
MHT	Mental Health Tribunal
MST	Multi-Systemic Therapy
MSU	Medium Secure Unit
NICE	National Institute for Health and Care Excellence
OCBs	Organisational Citizenship Behaviours

OM	Offender Manager
OPD	Offender Personality Disorder
OT	Occupational Therapist
PCAs	Pro-Criminal Attitudes
PCL-R	Psychopathy Checklist-Revised
PCT	Perceptual Control Theory
PICTS	Psychology Inventory of Criminal Thinking Styles
RC	Responsible Clinician
RNR	Risk-Need-Responsivity
SCT	Self-Categorisation Theory
SES	Socioeconomic Status
SI	Social Identity
SIP	Social Information Processing
SIT	Social Identity Theory
SPJ	Structured Professional Judgement
ST	Schema Therapy
TCV	Test for the Controlled Variable
TIC	Trauma Informed Care
VOTP	Violent Offenders Treatment Programme
VRP	Violence Reduction Programme
YOI	Young Offender Institution

1 Introduction

Joel Harvey and Derval Ambrose

> In correctional settings, where decisions have real consequences, it is tempting to think there is one "best" way of doing things. But this type of thinking gets us further, not closer, to our objectives as forensic psychologists. To be ethical and competent practitioners we need to be theoretically literate as well as technically competent.
>
> (Ward, 2020, p. 33)

Forensic practice is a complex interpersonal endeavour in which forensic practitioners, and the clients with whom they work, interact in multiple social contexts and relationships. Applied psychologists (forensic, clinical, counselling, educational, and organisational psychologists) work with clients who have been in contact with the criminal justice system (CJS) and they offer psychological assessments, formulations, interventions, training, and consultation to different services. Clients are referred to psychologists for a range of reasons, including risk to self, risk to others, personality difficulties, trauma, emotional dysregulation, psychosis, low mood, anxiety, and learning disabilities. At the centre of all forensic practice should be an attempt to understand the client's narrative, to make sense of their 'story', in order to work towards meaningful change (Rogers et al., 2015). Furthermore, there has been a growing awareness in forensic practice about the role of attachment and trauma in clients' histories (Rogers & Budd, 2015; Taylor et al., 2018; Willmot & Jones, 2022).

The contexts in which applied psychologists work are varied and include prisons, secure hospitals, youth offending teams, probation services, community forensic mental health teams, and various third sector organisations. Furthermore, applied psychologists work with clients on the edge of the CJS, such as pupils in schools, or with those who have an offending history but might only access generic mental health services rather than specific forensic ones. The relationship between mental health and punishment is complex (Taylor & Reeves, 2021). Clients move into, out of, and around multiple different systems, interacting with a multitude of professionals along the way, each of whom has their own cultural norms, language, and social identities. Often this leads to fragmented care and further disrupted attachments. Consider Dean's story in the following vignette:[1]

DOI: 10.4324/9781315560243-1

Dean is a 30-year-old man who is currently serving an indeterminate sentence for assaulting his partner. Dean grew up in inner-city London where he shared a three-bedroom local authority apartment with his mother, father, and two younger siblings. Dean's parents immigrated to the UK from Jamaica before Dean was born. Dean's father had an authoritarian presence in the household and frequently assaulted and belittled Dean's mother. His father taught Dean that women needed to be firmly managed in order to keep the family unit intact. Dean's father left the family home when Dean was 10 years old. His mother maintained three jobs as a cleaner to support her family. As a result, Dean cared for his younger siblings after school, and they were often alone late into the evening when his mother was at work. A neighbour reported that the children were alone for long periods of time, which subsequently prompted social services to become involved. Dean was taken into local authority care for nine months, during which time he was sexually assaulted by a member of staff in the care home.

Dean found school difficult, and frequently got into fights with other pupils. He was referred to local Child and Adolescent Mental Health Services (CAMHS) when he was 13 and received a diagnosis of Attention Deficit Hyperactivity Disorder (ADHD). Dean stopped attending appointments shortly afterwards. He was then moved to a Pupil Referral Unit (PRU) at the age of 13 and was permanently excluded from school when he was 15. Dean was exposed to multiple experiences of racism, on both personal and wider system levels. His mother always told him that this was the way the world was, and that he would need to work twice as hard as a white person to have any chance of success in life.

Dean became involved with a local gang when he was 12 years old. This gave him a sense of connection and a group of people he felt he could rely on. He was arrested for possessing cocaine when he was 14 years of age. He was on an Order with the Youth Offending Team (YOT) but found it difficult to attend appointments. This offence was followed by several other convictions over his teenage and young adult years, including possession of a weapon, assault, and burglary. Dean's offending continued throughout his life, and he has spent 10 of the past 15 years in custody at various points. During the periods of time that Dean was in the community he found it very difficult to secure employment due to his offending history.

It is evident from this brief account of Dean's story that, from a young age, he has inhabited varied social contexts and encountered a series of different professionals, each with their own agenda. Dean has experienced transitions between systems and has had to adapt to different settings in order to survive. While social relationships have been fragmented for Dean, and though they have left him feeling lonely and isolated, he has, at the same time, experienced varied social milieus and social encounters. We argue that understanding multiple contexts is key to the psychological endeavour, in order to enable and empower both clients and systems to bring about meaningful change.

As a result, the nature of forensic practice is inherently social and relational. To what extent do we as applied psychologists attend to the social, and make sense of it in relation to clinical forensic practice? Given the inherent social nature of forensic practice, in this book we argue for the value of applying a social psychological lens to practice. Social psychology, "the attempt to understand and explain how the thoughts, feelings, and behaviors of individuals are influenced by the actual, imagined, or implied presence of other human beings" (Allport, 1954, p. 3), has long been applied to the legal context. For example, it has been explored in relation to jury decision-making and key witness testimony (Helm, 2021; Slane & Dodson, 2022; Wixted et al., 2022). However, much less explicit attention has been paid to the potential practical value of key social psychological theories in forensic practice. This book explores that potential.

The Importance of Theory–Practice Connections

The need to make strong theory–practice connections lies at the heart of applied psychological practice (Jensen, 1999; Page et al., 2022). There has been a strong case made for applied psychologists to work as 'scientist-practitioners' (Jones & Mehr, 2007; Petersen, 2007) and to be trained to use the available evidence base to inform their practice. Applied psychologists are also taught to develop as 'reflective practitioners' (Lilienfeld & Basterfield, 2020), a skill that has been recognised as a core competency in applied psychological practice (Health and Care Professions Council, 2015). It is important for practising psychologists to reflect on their own emotional responses to their work, and their use of theory within practice.

Despite the developing connections between theory and practice within applied psychology training, it has been argued that, within forensic practice, we can be confronted with 'theoretical illiteracy' (Ward, 2020). Ward points out that theory can be used for conceptual analysis, and to predict, explain, and classify a phenomenon. In this book, when considering the application of social psychological theory, we do so at the 'explanatory' level; that is, a level at which the "theory seeks to account for the occurrence of phenomena (e.g., sexual arousal to children) by explaining why they exist and how they are instantiated" (ibid, p. 23). Similarly, we concur with Ward and advocate for "epistemic pluralism" (ibid, p. 28) in the application of theory and argue against "dogmatic acceptance" (ibid, p. 28) of existing theories.

In forensic settings, different practice frameworks are evident, and we are keen to explore the potential role of explicitly considering social psychological thinking in this variegated terrain. Ward and Durrant state:

a practice framework is a type of bridging theory that has been overlooked by researchers in the criminal justice area (as opposed to explanatory, treatment, and descriptive theories); one that guides key aspects of intervention from a specific perspective. It creates a bridge between treatment theories, normative assumptions, and etiological theories.

(Ward & Durrant, 2021, p. 2)

Practice frameworks "are hybrid, eclectic and there is no rule book for their construction even if one has a rough idea of what kinds of elements go into them" (Sullivan, 2022, p. 2). The aim of this book is not to develop a new practice framework but instead to highlight how social psychological theories and constructs might be used within existing practice frameworks, so that key concepts can be considered in a hybrid manner alongside other psychological processes.

Some interventions in forensic practice already draw upon social psychological constructs, such as attitude formation (Brown, 2010). For example, the role of offence-supportive *attitudes* is readily explored in programmes which address fire-setting (Gannon et al., 2012), sexual offending (Helmus et al., 2013), and violence reduction (VRP; Wong & Gordon, 2013). The literature on desistence from offending strongly focusses on *social identity* (Maruna, 2001; Mullins & Kirkwood, 2021), and the Good Lives Model (GLM; Ward, 2002; Ward & Brown, 2004)- is a strength-based practice framework which explicitly includes relatedness and community as core fundamental human needs. In this book, we would like to widen the net further and explore how forensic practitioners might apply a broader range of social psychological theories in their practice. Psychologists in forensic practice already draw on a range of social psychological theories and concepts in *implicit* ways but they do not always articulate them *explicitly*. For example, the role of *cognitive bias* has been explicitly explored in relation to forensic psychologists' mental health evaluations (Neal & Brodsky, 2016; Neal et al., 2022), but there is room for a broader consideration of social psychological concepts across different practices.

This book therefore examines more explicitly a range of social psychological theories and concepts and the ways that they are, or could be, applied to forensic psychological practice. We argue in this book that these theories are valuable when carrying out assessments (including risk assessments), formulations, and interventions with clients in forensic settings. These theories are also of direct relevance to psychological consultation, training, and the development of services. Moreover, these theories are valuable for understanding multi-disciplinary and multi-agency working, staff-client relationships, peer-to-peer relationships, and the public and political perception of the 'offender'. Furthermore, a social psychological lens enhances our understanding of systemic oppression and social injustice.

The aims are as follows:

1. To examine a range of social psychological theories and concepts which can be used to illuminate forensic practice.
2. To examine how these ideas from social psychology can aid practitioners' understanding of clients and the systems around them.
3. To examine how the application of social psychology can help psychologists to develop as scientist-practitioners and as reflective practitioners.
4. To encourage further theoretically-informed improvements in service delivery.

Having outlined the aims of the book, and before developing further the notion of applying social psychological thinking to forensic practice, we would first like to consider the role of the applied psychologist in social psychological research itself.

Imagining the Applied Psychologist's Role in Social Psychology

Social psychology has a long history of examining topics that are of relevance to forensic practice. For example, the Milgram (1965) study of obedience and the Stanford Prison Experiment (Haney et al., 1973) have long been established as core social psychological studies. However, it is important to note that ethical controversy surrounds these studies, and academics continue to debate their findings (Reicher & Haslam, 2006; Reicher & Haslam 2011; Reicher et al., 2020; Scott-Bottoms, 2020). The Stanford Prison Experiment (SPE), carried out in the 1970s in the basement of the psychology building at Stanford University, involved volunteer students being randomly assigned the role of 'prisoners' or 'guards'. However, the study had to be stopped after just six days due to the level of emotional distress experienced by the 'prisoners', and aggressive behaviour displayed by the 'guards'. The SPE highlights the complexities of power and control imbalances within forensic systems, and how this can manifest itself through individuals and groups.

When reflecting on the SPE (Haney et al., 1973) it is an interesting thought-experiment to imagine what role an applied psychologist would play if involved as a participant in the scenario. That is, it is interesting not to imagine a psychologist overseeing the ethics of the research study, but imagining participants being 'assigned' the role of a psychologist in the experiment's 'prison'. Indeed, in a real-world prison, prisoners and prison officers do not exist in a social vacuum. On the contrary, there are numerous other 'actors' on the 'stage' of a prison in the real world. In the UK, there are both psychologists employed by HM Prison Service and psychologists employed by the National Health Service (NHS) or private providers, who work in the prison as part of a mental health in-reach team, alongside substance misuse workers, members of the chaplaincy service, and other third-party professionals. But for the purpose of our imagined scenario here, let us pretend that Zimbardo assigned the role of applied psychologist to one of the students who participated in the research study. What role would the 'psychologist' play, and how would they respond? The participant's response to this hypothetical scenario might depend on what they are told about the role: do they see their role as supporting prisoner well-being or reducing risk and re-offending? And how would the psychologist be informed about who they work for within the structure of the 'prison' in the research study? Are they working for the prison? Are they an independent provider of services? This imagined scenario encourages us to reflect on the very essence of the role of a psychologist in a forensic setting, and the role-conflict that can emerge. How would a participant allocated the role of 'psychologist' in our imaginary 'prison experiment' see their identity? Moreover, what role might the experiment's designer have given that they already take leadership in the experiment (Haslam & Reicher, 2012; Haslam et al., 2019)? It would be of interest to consider whether the participants who assume the position of 'psychologist' would gravitate towards the role of 'officer' or 'prisoner', or indeed towards a different stance entirely. This question leads us to reflect more broadly on the social identity of psychologists. These questions continue to be at the centre of 'real-world' practice.

Applied psychologists working in real-world prisons and other secure settings occupy an interesting role in those spaces. The psychologists carry with them keys and, by virtue of doing so, also carry with them power, which can be enacted in different ways. For example, psychologists carry out risk assessments, which include determining whether someone should be recommended for release, and decide whether a client is ready to be part of a rehabilitation group, which is often needed for progression to the community. As part of a multi-disciplinary team, psychologists in forensic settings are asked to decide whether a patient detained under the Mental Health Act (1983), amended in 2007, should be granted leave or be able to 'step down' to a lower level of security. Psychologists as a social group hold power over a subordinate yet statistically larger group.

Yet applied psychologists have multiple clients: the prisoner or patient, the system (be that the prison or secure hospital setting), and the general public. In the real-world, psychologists hold a multifaceted identity, whereby certain aspects might become more salient within certain contexts. For example, a forensic psychologist might position themselves as someone who is there to support, while simultaneously be arguing to restrict someone's liberty. From the perspective of the prisoners, this may, understandably, not be experienced as 'care'. Psychologists in the prison system hold such complex roles that prisoners often have differing views about them. The identity of the psychologist might also depend on what other groups are present in the social world of the prison: does their identity shift when working with a psychiatrist who happens to subscribe to a purely medical model or with prison officers who might have an agenda of maintaining order? The psychologist might feel conflicted between recommending more restrictions (and hence punishment) or advocating for fewer restrictions and more freedom. The lines between 'care and collusion' and between 'boundary setting and punishing' are complex.

Ultimately, given the power-laden and low-trust environments of forensic settings, the applied psychologist needs to be self-aware about their position of power within the system and of the extent to, or moments in which, competing pressures could affect their sense of identity and role purpose. For example, in response to conflicting pressures, psychologists may feel inclined to be 'punitive' in their decision-making or 'resistant' against state power (actual or perceived 'collusion'). There is a tension here that is played out across different contexts. It unfolds during everyday interactions and can shape the culture or context in which psychologists work. This, in turn, could reciprocally shape their identity. Indeed, applied psychologists need to be aware of how they both shape, and are shaped by, the contexts in which they work.

The Role of Social Psychology in Forensic Practice

Given the social complexities revealed by the imagined scenario of the applied psychologist as a participant in a social psychological experiment, what might be the role of social psychological thinking in forensic practice? As noted earlier, our endeavour in this volume is not to produce a new practice framework, but to reflect on the value that social psychological thinking might bring to current frameworks. When considering the potential role of social psychological theories, it is important to first outline some of the therapeutic modalities and psychological frameworks that prevail in forensic practice. Applied psychological practice in forensic settings draws on a range of psychological therapies and overarching frameworks, such as the following (which are not an exclusive list):

- Cognitive Behavioural Therapy (CBT; Beck, 1967; Tafrate et al., 2018).
- Cognitive Analytic Therapy (CAT; Pollock et al., 2006; Ryle & Kerr, 2020).
- Multi-Systemic Therapy (MST; Fonagy et al., 2020; Henggeler et al., 1986).
- Dialectical Behaviour Therapy (DBT; Bianchini et al., 2019; Linehan, 1993).
- Schema-Focused Therapy (Bernstein et al., 2007; Young et al., 2006).

A range of practice frameworks exist, and the most common ones include:

- Risk Need Responsivity (RNR; offence-focused work; Andrews & Bonta, 2003). This model has three central tenets: treatment intensity should be matched to level of risk; treatment targets must be those areas that are related to the risk of offending ('criminogenic needs'); and interventions must be delivered in a way that is suitable to the needs of the recipient.

- The Good Lives Model (GLM; Ward & Maruna, 2007). This is a strengths-based rehabilitation framework that holds central the idea of promoting personal goals, while also addressing offending. The model focuses on developing resources and capabilities needed to achieve personal goals in pro-social ways.
- Power Threat Meaning (Johnstone et al., 2019). This framework explores what happened to an individual, how it affected them, what sense they made of it, and what they needed to do in order to survive. Additionally, the model explores the individual's strengths, and attempts to formulate their personal narrative.
- Trauma-Informed Practice (Willmot & Jones, 2022). Across prison, hospital, and community forensic teams, trauma-informed practice is gathering momentum. The key principles are safety, trust, choice, empowerment, and collaboration (see McGuire et al., 2022; Substance Abuse and Mental Health Services Administration, 2014).
- Restorative Justice (Braithwaite, 2002; Drennan, 2018; Moore, 2022). Restorative Justice (RJ) has been increasingly applied in forensic settings. A core tenet of this framework is 're-integrative shaming', whereby the wrongdoer works to take responsibility and reparation with the victim in a non-stigmatising context.

Within forensic practice in prisons, it is apparent that there is an artificial divide between 'offence-focused work' and support for mental health difficulties. Practically, offence-focused work is part of a prisoner's sentence planning and is supplied predominantly by forensic psychologists who are employed by HM Prison Service. Although forensic psychologists do also offer mental health support for prisoners, that is often instead the role of applied psychologists employed by the NHS Trust who provide mental health services within prisons. Within forensic hospital settings in the UK (low, medium, and high secure settings), applied psychologists work as part of multi-disciplinary teams to provide both offence-focused and mental health support for clients in their care.

In recent decades, rehabilitation in UK prisons has been dominated by the RNR model and CBT approaches. This has largely emerged from the 'What Works?' debate which explored the evidence base for effective interventions with people who have offended (Bonta & Andrews, 2017). As mentioned above, the RNR model has three central principles. The needs principle is the recommendation that treatment should directly target areas that have been empirically demonstrated to increase the risk of offending. Hence, for example, if an individual has a substance misuse history that is related to their offending, they would be recommended to complete a substance misuse intervention. Similarly, if they have committed a violent offence, they would be required to complete a violence reduction group intervention. However, the RNR model is a limited lens and does not allow for the inclusion of more inter-relational social factors. For example, the contexts of a traumatised history, psychological distress, and low self-esteem are not identified as being intervention targets.

This need to address wider contexts explains what a social psychological lens might bring to forensic practice and the existing practice frameworks named above. To help reflect on this, let's return to Dean, who was introduced earlier:

> During Dean's first year in custody, as part of his current sentence, he spent long periods of time in segregation due to assaulting other prisoners. He has been assigned a recently qualified forensic psychologist who is keen to ensure that he completes offence-focused group work. Dean does not agree with this and says that he was only managing his relationship in a way that was in keeping with his upbringing. The psychologist has been clear that he is unlikely to be successful at a Parole Board if he continues to maintain this attitude. She is also concerned that Dean is spending increasing amounts of time with prisoners who have known extremist views. In her most recent risk assessment of Dean, she concluded that he presents a high risk to the public, has psychopathic traits, and is at risk of being radicalised in prison.
>
> Recently, Dean attempted to hang himself in his cell and was referred to the mental health in-reach team. The team concluded that a referral to a specialist personality disorder service would be appropriate and helpful for Dean. The in-reach psychiatrist assessed Dean and concluded that he is 'malingering'. Furthermore, the team suggested that a personality disorder service would not be appropriate, as his personality traits cannot be changed. They have suggested that a low dose anti-psychotic medication could help Dean to manage his agitation, to return to regular location, and to engage in offence-focused group work.

Dean's story shows the complexity of the social contexts of forensic practice and the competing social systems that can impact on a person's journey through forensic environments. It also suggests some social psychological concepts and theories that could aid the practitioner in understanding Dean's journey and narrative. Dean has experienced a lot of *ostracism* in his life and this ostracism encouraged him to gravitate towards *groups* such as local gangs which likely helped him to feel connected and valued. Something similar may also have played out in a prison environment where he was also drawn to other prison gangs. The multiple traumas of exposure to violence, chronic racism, poverty, and gender *stereotyping* are likely to have had significant impact on his *attitudes* and *social identity*. Importantly, these experiences also mean that the suggestion that he join an offence-focused *group programme* is, at best, premature. It would be of interest to explore processes of *social influence* at play in Dean's life, and what influence the system is having on his sense of self and his behavioural responses.

Additionally, the vignette reveals the complex systems surrounding Dean and their impact on his experience of incarceration. As alluded to above, the forensic psychologist, as an employee of the Prison Service, arguably has multiple clients, including not only Dean himself but also the organisation and the public. In this situation, who should be prioritised? The artificial divide between the psychologist focused on risk assessment and offence-focused work and the mental health in-reach psychologist can be stark and lead to competing recommendations for Dean. Furthermore, it would also be of interest to examine the *attributions* that are made by Dean but also by professionals about Dean. And how does Dean need to present himself in order to survive across different settings? Indeed, what *impression management* strategies does he adopt, and how are these interpreted by professionals? We put forward three main arguments why social psychological thinking can be useful in our theoretical toolkit.

Firstly, we argue that a social psychological lens can shine a light on the extent to which existing therapeutic modalities consider the role of context. It allows us to view the social psychological and contextual thinking already at play within practice (albeit often implicitly), and the extent to which frameworks of practice keep context in mind. Essentially, a social psychological lens puts context at the centre of our thinking. Social psychology is a field of understanding which prioritises the association between individuals and the contexts in which social relations are enacted. With a social psychological perspective, we are required to keep context at the heart of our enquiry. This is important because, as Reicher and Haslam (2013) state, conversely, "by ignoring context, then, one necessarily ignores both how our understandings and behaviours are shaped by society and also how we can act to reshape society" (p. 114).

As detailed above, the current dominance of CBT interventions within an RNR framework focuses on so-called 'criminogenic needs'. We argue that this framework largely lacks an understanding of context and thus limits the psychologist's ability to understand the client's story. Other frameworks, such as trauma-informed practice, acknowledge that the relational context shapes the experience of clients and consider how the current environment can function to reinforce previous traumatic experiences. Indeed, when clients enter prison or other secure settings, they do not start from a neutral position-the environment is itself a 'threat', and each person is required to draw on whatever internal and external resources they can in order to survive.

Harvey and Smedley (2010) argued that a contextual understanding is needed because it allows psychologists to appreciate how the current experience of imprisonment (or other forensic settings) contributes to psychological difficulties. In doing so, psychologists are able to stay alert to the power imbalance in their practice, to understand the barriers to therapeutic engagement, to remember that prisons are low-trust environments that can affect therapy, and to ensure that they as psychologists do not rely solely on a pathological model of behaviour. It has been proposed elsewhere that applied psychologists would benefit from prison ethnography, both reading its results and following its process, as a way to become attuned to the contexts in which they practice. Indeed, thinking as an 'ethnographic practitioner' has been put forward as a way to ensure the psychologist is alert to the pains of imprisonment and able to take a reflective stance to the social milieu within which they too are constrained, but within which they contribute to the formation of the very structures that constrain themselves and others (Harvey, 2015). We argue that by taking a social psychological lens practitioners can expand their assessments, formulations, and interventions further to consider the context that shapes the person's journey into offending, and that serves to maintain or protect the person from offending and experiencing distress. This context of course includes the multiple systems that comprise the CJS.

Secondly, applied psychologists need to be aware of the potential unconscious processes that may be underscoring their choices and decision-making. For example, psychologists need to consider attributions that they and others make about behaviour when working as part of a multi-disciplinary team. Moreover, as imagined in the hypothetical scenario of the psychologist as a participant in Zimbardo's experiment, what social psychological factors might impact upon the psychologist's emotional and behavioural responses? What about their sense of self and social identity? A social psychological lens further highlights the need for reflection by the psychologist on their own social formation and for consideration of that during supervision. We argue that a social psychological lens is an additional tool to aid reflection both in and on practice.

Thirdly, explicitly adding a social psychological lens allows psychologists to reflect on their role as possible agents of change. Social psychologists have highlighted how "social identity theorising was not only concerned with the social nature of the self process, but also with the centrality of the social self to processes of change" (Reicher & Haslam, 2013, p. 115). A social psychological perspective encourages thinking about the relationship between the individual and the social structures they inhabit; it is of central importance to examine the relationship between structure and agency. It is important too to consider how psychologists might change the structures around them. Indeed, an interactionalist perspective is key here.

Outline of the Book

Chapter 2 considers how people explain events in their lives by causal attribution. The chapter considers key attribution errors, biases, and heuristics, including *fundamental attribution error*, *actor-observer bias*, and *self-serving attribution bias*, and the ways in which they might impact upon practice. The chapter examines how attributions affect how professionals make decisions about clients. Bowden, Glorney, and Durber provide practical recommendations for forensic assessment and treatment.

In Chapter 3, Morris and Webb explore *social identity theory* (Tajfel & Turner, 1986) and *self-categorisation theory* (Turner et al., 1987) and consider the significance of these theories to forensic psychological practice. The self does not exist in isolation from others; how we see ourselves is bound up in the groups with which we identify. The Criminal Social Identity model (CSI model; Boduszek & Hyland, 2011) and the Integrated Psychosocial Model of Criminal Social Identity (IPM-CSI; Boduszek et al., 2016) are explored and critiqued. The authors argue that it is important to develop a model of social identity that focuses on 'what happened to you', rather than using a model of 'criminal social identity' that infers that something is 'wrong with you'. The role of trauma and alienation and their relationship to 'disenfranchised' social identities are also explored.

Chapter 4 examines the role of impression management in forensic settings. Harvey and Drake draw on the work of Goffman (1990 [1959]) and reflect that both forensic clients and forensic practitioners are engaged in a 'self-presentation dance'. They argue that part of survival for forensic clients is to learn how to adapt their self-presentation differentially, depending on the audience. The chapter examines the idea of the 'public' and the 'private' self (Tice, 1992), as well as *self-discrepancy theory* which explores the discrepancies between the 'actual' and 'ideal' self (Higgins, 1987). The experience of shame and guilt is examined in relation to impression management, and the authors outline how to acknowledge this in practice. The chapter ends by considering practitioners' own impression management in their practice and argues that acknowledgement of contexts lies at the centre of effective practice.

In Chapter 5, Lara Arsuffi examines attitudes and the important role they play in our evaluations of the social world and interpersonal relationships. She explores attitude formation and the role of attitudes in relation to aggression, and she examines how attitudes are a key focus in offending behaviour programmes. The chapter also reflects on *social learning* and the *mere exposure effect* (Zajonc, 1968) and it considers how people who have offended might form and shape attitudes and personality when exposed to higher rates of abuse and neglect in their early lives. The chapter explores *cognitive dissonance* (Festinger, 1957) in relation to achieving change and ends by examining schemas and personality difficulties (Young et al., 2003).

Bruce and Rosenberger, in Chapter 6, focus on a social psychological understanding of aggression. The chapter outlines the definitions, typologies, and theories of aggression and then explores assessments and interventions. The authors put forward an adapted version of the Dynamic Appraisal of Situational Aggression (DASA; Ogloff & Daffern, 2006), in order to help work with clients displaying aggressive behaviour within forensic mental health settings.

In Chapter 7, Ambrose and Tancred examine groups in forensic practice. The chapter outlines the stages of *group formation* (Tuckman, 1965) and *group socialisation* over time (Moreland & Levine, 1982, 1984) and explores how these key concepts might improve group delivery in forensic settings. The authors also consider how concepts of *groupthink* (Janis, 1971, 1972, 1982) and *group polarisation* (Isenberg, 1986) could have relevance to groups of practitioners making decisions about risk and progression.

Huddy and Carey examine *coercion and control* within forensic settings, in Chapter 8. They highlight Perceptual Control Theory (PCT; Powers, 1973, 1998) and its relationship to social interactions. They emphasise that recipients of treatment interventions may well experience those interventions as coercion. They also discuss the inevitability of separate perspectives across many staff groups and the importance of keeping sight of the perspective of the service user. Finally, the authors explore the potential application of the PCT perspective in order to improve forensic practice implementation in a way that minimises coercion.

In Chapter 9, Kaip and Harvey examine research on *ostracism* (Williams, 2007; Zadro et al., 2004). The temporal need-threat model of ostracism (Williams, 2009) is explored in relation to forensic practice. In the chapter, it is argued that forensic practitioners need to consider the potentially harmful impact of social ostracism on their clients. It is argued that ostracism is an important construct to consider when assessing clients, and developing psychological formulations. Ostracism is considered in relation to violence, the experience of imprisonment, and suicidality. Through composite case studies, Kaip and Harvey argue for the inclusion of social exclusion in forensic practice.

The final chapter extensively sets out the role of *stereotyping and prejudice* in forensic practice. Ambrose, Campbell, and Kaip consider the impact of stereotyping and prejudice across three broad areas of race and ethnicity, gender, and class. They propose how a model of ethnic identity (Phinney, 1989) might be applied to therapeutic work with people in contact with the CJS, and consider the importance of being a culturally competent practitioner. Additionally, they examine how *gender role strain* (Pleck, 1981, 1995), the fundamental attribution error (Ross, 1977), and gender responsive environments are important in forensic practice. Finally, the authors explore the social psychology of class and class-based stereotyping and prejudice in criminal justice settings. Concepts of intersectionality and multiple experiences of oppression are key to this chapter.

Conclusion

In clinical forensic practice, there are examples of professionals thinking of ways to address the power imbalance between groups. There is an emerging discourse of working collaboratively with clients to consider the 'least strict practices' (Sustere & Tarpey, 2019), to develop clients' knowledge of the tools used, to increase peer involvement in staff recruitment interviews, and to consider peer mentoring. While there is an acknowledgement of the power imbalance within forensic practice, how far does this really take us? If we were to ask Dean, would he feel that he was being 'rehabilitated' or that he was being 'coerced' (Simms-Sawyers et al., 2020)? The power imbalance continues to exist, and we need to be aware of the notion of coercion when patients engage in therapy. So is the discourse surrounding collaboration actually coercion in disguise? This question is especially important given that most clients come from backgrounds which were traumatic, had low opportunities, and/or were stigmatised.

In acknowledging these adverse experiences of forensic clients, we should consider the role of a psychologist in the wider social systems that generate these experiences. Is it to 'rehabilitate' or 'treat' the *individual* client, to reduce their risk to the public, or to reduce their psychological distress? Or is it to effect change in the wider social *systems* that perpetuate trauma and abuse? Perhaps, the point is that psychologists have responsibilities at both individual and systemic levels; they need not only to acknowledge systemic factors but to advocate for their change.

We would argue that it is also the role of applied psychologists to be reflective about when to be mobilisers of change, to readdress some of the systemic difficulties both within and outside of forensic institutions. The concept of community psychology, which we would argue has a social psychological lens, is central here. Community psychology is a strengths-based approach that emerged in the USA in the 1960s during times of protests against the Vietnam war and in support of the Civil Rights Movement (Jason et al., 2019). The goals of community psychology have been to "examine and better understand complex individual-environment interactions in order to bring about social change, particularly for those who have limited resources and opportunities" (Jason et al., 2019, p. 5). The approach of community psychology moves away from notions of simply treating the individual and instead moves toward achieving larger group and systemic changes (ibid.). This is not dissimilar to a Public Health model that attempts to enact change through prevention, such as immunisation programmes or taxing tobacco products. Inequalities throughout social structures are known to impact and cause many health and social difficulties (Wilkinson & Picketts, 2010). As such, any effective reduction in these difficulties would involve social change and the redistribution of power (Albee, 1986). We argue that the same premise exists for reducing crime and psychological distress.

Another key consideration for applied psychologists in their practice is intersectionality. The term 'intersectionality' was first introduced by Crenshaw in 1989 and has its origins in the work of feminist and critical theorists. Intersectionality describes a systemic approach to exploring the meaning and consequences associated with membership of various social groups (Crenshaw, 1989). Collins (2000) proposes that "intersectional paradigms suggest that certain ideas and/or practices surface repeatedly across multiple systems of oppression" (Collins, 2000, p. 47). However, the extent to which psychology has integrated this theory is questionable. Indeed, Settles et al. (2020) argue that resistance to the theory in psychology represents an epistemic exclusion, which they describe as "the devaluation of some scholarship as illegitimate and certain scholars as lacking credibility" (p. 1). They postulate that the reason for this exclusion is that this theory challenges the most dominant psychological norms, and they argue that this epistemic exclusion occurs in both formal ways (such as omitting discussion of this topic from mainstream journals) and more informal ones (such as misrepresenting the theory; Settles et al., 2020). Furthermore, they conclude that such exclusion creates a significant obstacle to social justice in psychology.

We argue that making social psychological thinking more central to applied forensic psychological practice would ensure that systemic injustice would be central to our thinking when we attempted to understand the stories of our clients. In fact, the role of the applied psychologist is not only to ensure that there is an understanding of social injustice in our assessments, formulations, and interventions, but in our role proactively and carefully to challenge that social injustice. Indeed, it is important for psychologists to reflect on how their roles have been conceptualised and positioned within the system, and to consider their own agency in relation to changing the social injustice that pervades the lives of many clients.

Social psychology is not here argued to be another 'answer' to all questions in psychological practice. Indeed, we do not think there is "one 'best' way of doing things" (Ward, 2020, p. 33), despite the claims made for randomised control trials. However, we hope that this volume, by connecting forensic practice and social psychology more explicitly might enable further appreciation of the stories of our clients, the complex role of the multiple contexts they live in, and indeed the roles and contexts of applied psychologists themselves.

Note

1 In this book, all vignettes are composite. The vignettes are based on the clinical forensic experience of authors working with different clients and have been developed so that no one client can be identified.

References

Albee, G. W. (1986). Toward a just society: Lessons from observations on the primary prevention of psychopathology. *American Psychologist*, 41(8), 891–898.

Allport, G. (1954). The historical background of modern social psychology. In G. Lindzey (ed.), *Handbook of social psychology* (pp. 3–56). Addison Wesley.

Andrews, D. A., & Bonta, J. (2003). *The psychology of criminal conduct* (3rd edition). Anderson.

Beck, A. T. (1967). *Depression: Causes and treatment*. University of Pennsylvania Press.

Bernstein, D. P., Arntz, A., & de Vos, M. (2007). Schema focused therapy in forensic settings: Theoretical model and recommendations for best clinical practice. *International Journal of Forensic Mental Health*, 6(2), 169–183.

Bianchini, V., Cofini, V., Curto, M., Lagrotteria, B., Manzi, A., Navari, S., Ortenzi, R., Paoletti, G., Pompili, E., Pompili, P. M., Silvestrini, C., & Nicolò, G. (2019). Dialectical behaviour therapy (DBT) for forensic psychiatric patients: An Italian pilot study. *Criminal Behaviour and Mental Health*, 29(2), 122–130.

Boduszek, D., & Hyland, P. (2011). The theoretical model of criminal social identity: Psychosocial perspective. *International Journal of Criminology and Sociological Theory*, 4(1), 604–614.

Boduszek, D., Dhingra, K., & Debowska, A. (2016). The Intergrated Psychosocial Model of Criminal Social Identity (IPM-CSI). *Deviant Behavior*, 37(9), 1023–1031.

Bonta, J., & Andrews, D. A. (2017). *The psychology of criminal conduct* (6th edition). Routledge.

Braithwaite, J. (2002). *Restorative justice and responsive regulation*. Oxford University Press.

Brown, J. (2010). Social psychological theories applied to forensic psychology topics. In J. M. Brown & E. A. Campbell (eds), *The Cambridge handbook of forensic psychology* (pp. 111–117). Cambridge University Press.

Collins, P. H. (2000). Gender, Black Feminism, and Black Political Economy. *The Annals of the American Academy of Political and Social Science*, 568, 41–53.

Crenshaw, K. (1989). Demarginalizing the intersection of race and sex: a Black feminist critique of anti-discrimination doctrine, feminist theory and anti-racist politics. *University of Chicago Legal Forum*, 1989(1), 139–167.

Drennan, G. (2018). Restorative justice applications in mental health settings- pathways to recovery and restitution. In J. Adlam, T. Kluttig, & B. X. Lee (eds), *Violent states and creative states: From the global to the individual, book II: Human violence and creative humanity* (pp. 181–194). Jessica Kingsley Publishers.

Festinger, L. (1957). *A theory of cognitive dissonance*. Stanford University Press.

Fonagy, P., Butler, S., Cottrell, D., Scott, S., Pilling, S., Eisler, I., Fuggle, P., Kraam, A., Byford, S., Wason, J., Smith, J. A., Anokhina, A., Ellison, R., Simes, E., Ganguli, P., Allison, E., & Goodyer, I. M. (2020). Multisystemic therapy versus management as usual in the treatment of adolescent antisocial behaviour (START): 5-year follow-up of a pragmatic, randomized, controlled, superiority trial. *The Lancet*, 7(5), 420–430.

Gannon, T., Ciardha, C. Ó., Doley, R., & Alleyne, E. (2012). The multi-trajectory theory of adult firesetting (M-TTAF). *Aggression and Violent Behavior*, 17, 107–121.

Goffman, E. (1990 [1959]). *The presentation of self in everyday life*. Penguin.

Haney, C., Banks, C., & Zimbardo, P. (1973). A study of prisoners and guards in a simulated prison. *Naval Research Reviews*, 9, 1–17.

Harvey, J. (2015). The ethnographic practitioner. In D. H. Drake, R. Earle & J. Sloan (eds), *The Palgrave handbook of prison ethnography* (pp. 390–402). Palgrave Macmillan.

Harvey, J. & Smedley, K. (2010). *Psychological therapy in prisons and other secure settings*. Willan Publishing.

Haslam S. A., & Reicher S. D. (2012). Contesting the 'nature' of conformity: What Milgram and Zimbardo's studies really show. *PLoS Biology*, 10(11), e1001426.

Haslam, S. A., Reicher, S. D., & Van Bavel, J. J. (2019). Rethinking the nature of cruelty: The role of identity leadership in the Stanford Prison Experiment. *American Psychologist*, 74, 809–822.

Health and Care Professions Council. (2015). Standards of proficiency – Practitioner psychologists. Retrieved from www.hcpc-uk.org/globalassets/resources/standards/standards-of-proficiency—practitioner-psychologists.pdf?v=637106257690000000.

Helm, R. K. (2021). Evaluating witness testimony: Juror knowledge, false memory, and the utility of evidence-based directions. *The International Journal of Evidence & Proof*, 25(4), 264–285.

Helmus, L., Hanson, R. K., Babchishin, K. M., & Mann, R. E. (2013). Attitudes supportive of sexual offending predict recidivism: A meta-analysis. *Trauma, Violence, & Abuse*, 14(1), 34–53.

Henggeler, S. W., Rodick, J. D., Borduin, C. M., Hanson, C. L., Watson, S. M., & Urey, J. R. (1986). Multisystemic treatment of juvenile offenders: Effects on adolescent behavior and family interactions. *Developmental Psychology*, 22, 132–141.

Higgins, E. T. (1987). Self-discrepancy: A theory relating self and affect. *Psychological Review*, 94, 319–340.

Isenberg, D. J. (1986). Group polarization: A critical review and meta-analysis. *Journal of Personality and Social Psychology*, 50(6), 1141–1151.

Janis, I. L. (1971). Groupthink. *Psychology Today*, November, 43–84.

Janis, I. L. (1972). *Victims of groupthink: A psychological study of foreign-policy decisions and fiascoes*. Houghton-Mifflin.

Janis, I. L. (1982). *Groupthink: Psychological studies of policy decisions and fiascoes* (2nd edition). Houghton-Mifflin.

Jason, L. A., Glantsman, O., O'Brien, J. F., & Ramian, K. N. (2019). Introduction to the field of Community Psychology. In L. A. Jason, O. Glantsman, J. F. O'Brien, & K. N. Ramian (eds), *Introduction to community psychology: Becoming an agent of change* (pp. 3–22). Rebus Press.

Jensen, P. S. (1999). Links among theory, research, and practice: Cornerstones of clinical scientific progress. *Journal of Clinical Child Psychology*, 28(4), 553–557.

Johnstone, L., Boyle, M., Cromby, J., Dillion, J., Harper, D., Kinderman, P., Longden, E., Pilgrim, D., & Read, J. (2019). Reflections on responses to the Power Threat Meaning Framework one year on. *Clinical Psychology Forum*, 313, 47–54.

Jones, J. L., & Mehr, S. L. (2007). Foundations and Assumptions of the Scientist-Practitioner Model. *American Behavioral Scientist*, 50(6), 766–771.

Lilienfeld S. O., & Basterfield, C. (2020). Reflective practice in clinical psychology: Reflections from basic psychological science. *Clinical Psychology: Science and Practice*, 27(4), e12352.

Linehan, M. M. (1993). Dialectical behavior therapy for treatment of borderline personality disorder: implications for the treatment of substance abuse. *NIDA Research Monograph*, 137, 201–201.

Maruna, S. (2001). *Making good: How ex-convicts reform and rebuild their lives*. American Psychological Association.

McGuire, F., Carlisle, J., & Clark, F. (2022). Trauma-informed care in secure psychiatric hospitals. In P. Willmot & L. Jones (eds), *Trauma-informed forensic practice* (pp. 348–362). Routledge.

Mental Health Act. (1987). Mental Health Act. Retrieved from https://cdn.nic.in/SJ/PDFFiles/Sparsh_mentalhealthact.pdf.

Milgram, S. (1965). Some conditions of obedience and disobedience to authority. *Human Relations*, 18(1), 57–76.

Moore, E. (2022). Trauma and restorative justice. In P. Willmot, & L. Jones (eds), *Trauma-informed forensic practice* (pp. 396–412). Routledge.

Moreland, R. L., & Levine, J. M. (1982). Socialisation in small groups: Temporal changes in individual-group relations. In L. Berkowitz (ed.), *Advances in experimental social psychology* (pp. 137–192). Academic Press.

Moreland, R. L., & Levine, J. M. (1984). Role transitions in small groups. In V. L. Allan, & E. Van De Vliert (eds), *Role transitions: Explorations and explanations* (pp. 181–195). Plenum.

Mullins, E., & Kirkwood, S. (2021). Co-authoring desistance narratives: Analysing interactions in groupwork for addressing sexual offending. *Criminology and Criminal Justice*, 21(3), 316–333.

Neal, T. M. S., & Brodsky, S. L. (2016). Forensic psychologists' perceptions of bias and potential correction strategies in forensic mental health evaluations. *Psychology, Public Policy, and Law*, 22(1), 58–76.

Neal, T. M. S., Lienert, P., Denne, E., & Singh, J. P. (2022). A general model of cognitive bias in human judgment and systematic review specific to forensic mental health. *Law and Human Behavior*, 46(2), 99–120.

Ogloff, J. R., & Daffern, M. (2006). The dynamic appraisal of situational aggression: an instrument to assess risk for imminent aggression in psychiatric inpatients. *Behavioral Sciences & the Law*, 24(6), 799–813.

Page, A., Stritzke, W. G. K., & McEvoy, P. M. (2022). *Clinical psychology for trainees: Foundations of science-informed practice* (3rd edition). Cambridge University Press.

Petersen, C. A. (2007). A Historical Look at Psychology and the Scientist-Practitioner Model. *American Behavioral Scientist*, 50(6), 758–765.

Phinney, J. S. (1989). Stages of ethnic identity development in minority group adolescents. *The Journal of Early Adolescence*, 9(1–2), 34–49.

Pleck, J. H. (1981). *The myth of masculinity*. MIT Press.

Pleck, J. H. (1995). The gender role strain paradigm: An update. In R. F. Levant & W. S. Pollack (eds), *A new psychology of men* (pp. 11–32). Basic Books.

Pollock, P. H., Stowell-Smith, M., & Göpfert, M. (2006). *Cognitive analytic therapy for offenders: A new approach to forensic psychotherapy*. Routledge.

Powers, W. T. (1973). *Behavior: The control of perception*. Aldine.

Powers, W. T. (1998). *Making sense of behavior*. Benchmark Publications.

Reicher, S., & Haslam, S. A. (2006). Rethinking the psychology of tyranny: The BBC Prison Study. *British Journal of Social Psychology*, 45, 1–40.

Reicher, S. D., & Haslam, S. A. (2011). After shock? Towards a social identity explanation of the Milgram 'obedience' studies. *British Journal of Social Psychology*, 50, 163–169.

Reicher, S., & Haslam, A. (2013). Towards a 'science of movement': Identity, authority and influence in the production of social stability and social change. *Journal of Social and Political Psychology*, 1(1), 112–131.

Reicher, S. D., van Bavel, J. J., & Haslam, S. A. (2020). Debate around leadership in the Stanford Prison Experiment: Reply to Zimbardo and Haney (2020) and Chan et al. (2020). *American Psychologist*, 75(3), 406–407.

Rogers, A., & Budd, M. (2015). Developing safe and strong foundations: The DART framework. In A. Rogers, J. Harvey, & H. Law (eds), *Young people in forensic mental health settings: Psychological thinking and practice* (pp. 356–389). Palgrave Macmillan.

Rogers, A., Harvey, J., Law, H., & Taylor, J. (2015). Introduction. In A. Rogers, J. Harvey, & H. Law (eds), *Young people in forensic mental health settings: Psychological thinking and practice* (pp. 1–19). Palgrave Macmillan.

Ross, L. D. (1977). The intuitive psychologist and his shortcomings: Distortions in the attributional process. In L. Berkowitz (ed.), *Advances in experimental social psychology* (pp. 173–220). Academic Press.

Ryle, A., & Kerr, I. B. (2020). *Introducing cognitive analytic therapy: Principles and practice of a relational approach to mental health* (2nd edition). Wiley.

Scott-Bottoms, S. (2020). The dirty work of the Stanford Prison Experiment: Re-reading the dramaturgy of coercion. *Incarceration: An International Journal of Imprisonment, Detention and Coercive Confinement*, 1(1), 1–188.

Settles, I. H., Warner, L. R., Buchanan, N. T., & Jones, M. K. (2020). Understanding psychology's resistance to intersectionality theory using a framework of epistemic exclusion and invisibility. *Journal of Social Issues*, 76(4), 796–813.

Simms-Sawyers, C., Miles, H., & Harvey, J. (2020). An exploration of perceived coercion into psychological assessment and treatment within a low secure forensic mental health service. *Psychiatry, Psychology and Law*, 27(4), 578–600.

Slane, C. R., & Dodson, C. S. (2022). Eyewitness confidence and mock juror decisions of guilt: A meta-analytic review. *Law and Human Behavior*, 46(1), 45–66.

Substance Abuse and Mental Health Services Administration. (2014). SAMHSA's concept of trauma and guidance for a trauma-informed approach. Retrieved from https://ncsacw.acf.hhs.gov/userfiles/files/SAMHSA_Trauma

Sullivan, J. (2022). The concept of practice frameworks in correctional psychology: A critical appraisal. *Aggression and Violent Behavior*, 63, 107116.

Sustere, E. & Tarpey, E. (2019). Least restrictive practice: its role in patient independence and recovery. *The Journal of Forensic Psychiatry & Psychology*, 30(4), 614–629.

Tafrate, R.C., Mitchell, D., & Simourd, D. J. (2018). *CBT with justice-involved clients: Interventions for antisocial and self-destructive behaviors*. Guildford Press.

Tajfel, H., & Turner, J. C. (1986). The social identity theory of intergroup behavior. In S. Worchel, & W. G. Austin (eds), *Psychology of intergroup relation* (pp. 7–24). Hall Publishers.

Taylor, P., & Reeves, A. (2021). Introduction. In P. Taylor, S. Morley, & J. Powell (eds), *Mental health and punishments: Critical perspectives in theory and practice* (pp. 1–5). Routledge.

Taylor, J., Shostak, L., Rogers, A., & Mitchell, P. (2018). Rethinking mental health provision in the secure estate for children and young people: A framework for integrated care (SECURE STAIRS). *Safer Communities*, 17(4). 193–201.

Tice, D. M. (1992). Self-concept change and self-presentation: The looking glass self is also a magnifying glass. *Journal of Personality and Social Psychology*, 63(3), 435–451.

Tuckman, B. W. (1965). Developmental sequence in small groups. *Psychological Bulletin*, 63(6), 384–399.

Turner, J. C., Hogg, M. A., Oakes, P. J., Reicher, S. D., & Wetherell, M. S. (1987). *Rediscovering the social group: A self-categorization theory*. Blackwell.

Ward, T. (2002). Good lives and the rehabilitation of offenders: Promises and problems. *Aggression and Violent Behavior*, 7(5), 513–528.

Ward, T. (2020). Why theory matters in correctional psychology. *Forensische Psychiatrie, Psychologie, Kriminologie*, 14, 22–34.

Ward, T., & Brown, M. (2004). The Good Lives Model and conceptual issues in offender rehabilitation. *Psychology, Crime & Law*, 10(3), 243–257.

Ward, T., & Durrant, R. (2021). Practice frameworks in correctional psychology: Translating causal theories and normative assumptions into practice. *Aggression and Violent Behavior*, 58, 101612.

Ward, T., & Maruna, S. (2007). *Rehabilitation: Beyond the risk paradigm*. Routledge.

Wilkinson, R., & Pickett, K. (2010). *The spirit level: Why equality is better for everyone*. Penguin.

Williams, K. D. (2007). Ostracism: The kiss of social death. *Social and Personality Psychology Compass*, 1(1), 236–247.

Williams, K. D. (2009). Ostracism: A temporal need-threat model. In M. Zanna (ed.), *Advances in experimental social psychology* (pp. 279–314). Academic Press.

Willmot, P., & Jones, L. (eds). (2022). *Trauma-informed forensic practice*. Routledge.

Wixted, J. T., Mickes, L., Brewin, C. R., & Andrews, B. (2022). Doing right by the eyewitness evidence: a response to Berkowitz et al. *Memory*, 30(1), 73–74.

Wong, S. C., & Gordon, A. (2013). The violence reduction programme: A treatment programme for violence-prone forensic clients. *Psychology, Crime & Law*, 19(5–6), 461–475.

Young, J. E., Klosko, J. S., & Weishaar, M. E. (2003). *Schema therapy*. Guilford Press.

Young, J. E., Klosko, J. S., & Weishaar, M. E. (2006). *Schema therapy: A practitioner's guide*. Guilford Press.

Zadro, L., Williams, K. D., & Richardson, R. (2004). How low can you go? Ostracism by a computer is sufficient to lower self-reported levels of belonging, control, self-esteem, and meaningful existence. *Journal of Experimental Social Psychology*, 40(4), 560–567.

Zajonc, R. B. (1968). Attitudinal effects of mere exposure. *Journal of Personality and Social Psychology*, 9(2), 1–27.

2 Attributions and Biases

Laura Bowden, Emily Glorney and Emily Durber

Introduction

Making sense of our own behaviour, the behaviour of others, and the world we live in, is something we all attempt, every day. Social psychology has long sought to study the process of assigning causality (attribution) to behaviour and events (Hogg & Vaughan, 2018) and understand the ways in which it can influence an individual's thoughts, feelings, and decision-making. In addition to studying attributions, research has also explored the mental processes involved in making judgements, including, for example, the tendency to gravitate towards or against certain information (i.e. biases), and what the implications of this might be (ibid.). However, a noticeable lack of attention has been paid to understanding what this knowledge means for the world of forensic practice (Neal et al., 2022). Are professionals who are involved in forensic risk assessment/treatment, for example, free from the biases that others hold? As forensic practitioners, should we be paying more attention to the biases that we bring into practice, and taking action to address these? These are just some of the questions and issues addressed in this chapter.

Throughout this chapter, the fictional case study of Antony, introduced below, will be used to illustrate the role of attributions and biases in forensic practice. We encourage you to think about what factors could influence how Antony progresses through the system.

> Antony has been transferred to a secure psychiatric hospital from prison following an assessment where he received a diagnosis of schizophrenia. He had a previous diagnosis of antisocial personality disorder with paranoid traits. Antony was originally convicted of attempted murder of a male associate, and received an Indeterminate Sentence for Public Protection with a minimum tariff of eight years. While in prison, Antony carried out two further acts of serious violence against a fellow prisoner and a prison officer, and was experiencing increasing paranoia and auditory and visual hallucinations. Following the most recent assaults, he was transferred to a hospital setting for assessment and treatment. In order to be discharged/released, Antony will undergo

DOI: 10.4324/9781315560243-2

assessment and treatment by a range of professionals in hospital. When his mental health has improved, he will be returned to prison to undergo further assessment of risk of reoffending, and a decision about his release will be made by the Parole Board of England and Wales.

(Context 1)

Understanding the way in which we form judgments, and the errors and biases that can occur with them, is essential in helping practitioners to reflect on the complex nature of forensic assessment and treatment, and become more aware of their intrinsic mechanisms for cognitive 'short cuts'. In the context of social psychological theory, this chapter will begin by exploring how people explain events in their own and others' lives through causal attribution, and attribution errors. By considering concepts such as implicit racial, cultural, and gender bias, this chapter will then explore how we form impressions of others. We will then examine literature relating to biases in forensic settings, and appraise the ways in which they can influence forensic practitioner decision-making. Finally, the chapter will conclude with a set of practical recommendations for forensic assessment and treatment. Throughout the chapter, we will consider how these concepts apply to the case of Antony as we follow his journey through the criminal justice system (CJS).

How Social Psychological Theory Contributes to How We Make Sense of Events in our Lives and Those of Others

Social psychological theory is just one domain that contributes to the research literature on how we make sense of events and the world around us. From a biopsychosocial perspective, there are factors that influence an individual's ability to generate an understanding that would be generally shared with other people. In other words, there are biological, neuropsychological, and cognitive factors that shape the lens through which meaning, and understanding, are made (Stainton Rogers, 2020). With acknowledgement of the breadth of factors that can contribute to how we make sense of ourselves and the world around us, this section will focus on social psychological theories. It will start with an overview of attribution theory, followed by an exploration of the relative contribution of biases.

Attribution Theory

During the latter half of the 1950s, when there was an upsurge of theoretical exploration into disciplines related to social psychology, Heider (1958) published what is now considered to be the principal source of attribution theory. The theory itself is concerned with how we generate explanations for events that we experience (e.g. our own and others' behaviour), and is rooted in the work of several authors. These include Kelley (1967) who introduced the covariation model which proposes that people assign the cause of behaviour to the factor that most closely covaries with it. In other words, people act like scientists to work out why someone is behaving in a particular way by assessing factors about the person and situation (Hogg & Vaughan, 2018). Additionally, Jones and Nisbett (1972) elaborated on several biases involved in attributions, and Weiner (1986) defined stability, locus, and control as the three main characteristics of attributions that can influence an individual's motivation to participate in that behaviour in the future. It is one of the many social psychological theories that has direct relevance to our everyday experiences. For example, it is linked to the questions we ask ourselves while driving ('why did they pull out in front of me?'), waiting to be served at a restaurant ('why is service taking so long?'), and watching the news ('what would make someone commit such a horrific crime?'). In response to each of these questions, we either make an internal (relating to personal factors considered to be within an individual's control, such as their personality or temperament) or external (relating to factors beyond a person's control, such as their circumstance or situation) attribution as to why the behaviour took place. So, it might be taking a while to get served in a restaurant because the server is not motivated to deliver good customer service (internal), or instead because the restaurant is much busier than the management anticipated, and there is a shortage of staff on shift (external).

Unbeknownst to students, attribution theory is likely one of the main reasons why they continue on to graduate study in psychology. In 1958, Heider proposed a theory of naïve or common-sense psychology; an idea most psychologists will have personal experience with when attempting to give lay explanations of complex behaviour to family and friends, only to be met with the response: 'well, that's just common sense!' Although the response might make one question the extensive training pathway they undertook to qualify as a psychologist, the response is evidence of Heider's theory. We all seek to understand what motivates the behaviour of other people, and why people behave in the way that they do, to have a common sense of understanding. This is frequently a driver for graduate study in psychology.

Not making sense of events or the behaviour of other people can create psychological discomfort. So, even when we do not have much information on which to generate a comprehensive understanding, we tend to make sense of behaviour as best we can and in a way that feels most psychologically comfortable. If we return to the question of 'what would make someone commit such a horrific crime?', it might be that, with the knowledge that we have of the event, we find ourselves saying 'I would do the same in that situation'. However, given that we labelled the crime as 'horrific', this assumption will likely make us feel psychologically uncomfortable. To reduce this discomfort, we will probably make an internal, rather than external, attribution about the event: 'there must be something wrong with that person; they are so different from me'. This is an example of cognitive dissonance (Cushman, 2019; Festinger, 1957); that is, we attribute motivation and cause to the events and behaviour of others in a way that helps us to make sense of the world around us, and protects how we think and feel about ourselves.

As practitioners within forensic services, we need to pay particular attention to the internal attributions that we make about others' behaviour. By the very nature of the people who are detained in, or engage with, forensic services (i.e. people who have transgressed a legal boundary or 'done something wrong', and/or people who have a mental disorder), we are primed to make inferences about internal factors that motivate behaviour.

Of course, in making sense of events and behaviour that have already occurred, we build up a body of experience that helps us to predict how people will behave in future situations. Applying learning to future contexts is helpful because it enables us to quickly appraise a situation, and informs our responses in a way that best reduces psychological discomfort. Cognitive shortcuts (i.e. heuristics), of which bias is a by-product, are functional and useful; they aid the speed and efficiency with which we arrive at solutions (Tversky & Kahneman, 1974). However, while heuristics help us to quickly make generally adaptive decisions with limited demand on our cognitive resource, they can become problematic when informed by our error-prone, automatic interpretations of events. It is therefore when we systematically under- or overuse available information to make inaccurate inferences about behaviour that *attribution bias* occurs (Hogg & Vaughan, 2018). Consider Antony again in Context 2:

> Antony has been placed in a challenging ward environment in the secure hospital. There are a number of patients who have been unsettled (responding to unknown stimuli and becoming aggressive towards peers on the ward), and this has resulted in multiple altercations and restraints. Antony has a history of attachment trauma. His father left the family home when he was a baby, and his stepfather was regularly violent towards him, his siblings, and his mother. The secure hospital living environment is exceptionally difficult for Antony to cope with, and he has become increasingly 'argumentative' and 'rude' to staff. In the multidisciplinary team meeting, the professionals discuss this recent behaviour as further evidence of personality disorder. Antony's primary nurse, who has been on bias awareness training, raised the question of whether this behaviour is evidence of personality disorder or more to do with the ward situation. She then asked how the other team members would behave if they were living on the ward with the current instability.
>
> (Context 2)

According to Jones and Davis (1965), we are more likely to make internal attributions about behaviour when we make a link between *it* and an internal motivation. In Antony's case, as illustrated above, his multidisciplinary team are aware of his antisocial personality diagnosis, and link this internal driver to his argumentative and rude behaviour. In doing so, they are making a correspondence inference (Jones & Davis, 1965); that is, believing that

Antony was intentionally rude and argumentative because his behaviour corresponds to maladaptive dispositional personality traits. Making a correspondence inference here is problematic because it discounts other potential explanations for Antony's behaviour.

A bias in making correspondence inferences reflects a tendency to attribute stable personality characteristics to behaviour. If you are familiar with the fundamental attribution error (Ross et al., 1977) then you might see a similarity here; the term is used interchangeably with correspondence bias but reflects a tendency to make internal attributions *even* with the knowledge of clear external causes. For both correspondence bias and fundamental attribution error, a problem is the lack of consideration of external attribution; you can appreciate how this could be problematic for Antony, and an example of this in practice is presented in Figure 2.1 later in this chapter.

In Context 2, Antony's primary nurse challenges the correspondence bias made by the multidisciplinary team by inviting them to put themselves in Antony's situation, and consider how they would behave. Some team members might think 'yes, I would probably do the same or something similar in that situation. It must be really challenging to be in this environment in the context of the experiences Antony has had throughout his life'. In inviting team members to consider an alternative explanation, Antony's primary nurse could be challenging the actor-observer effect (Jones & Nisbett, 1972); the tendency to attribute internal causes (e.g. disposition) to the

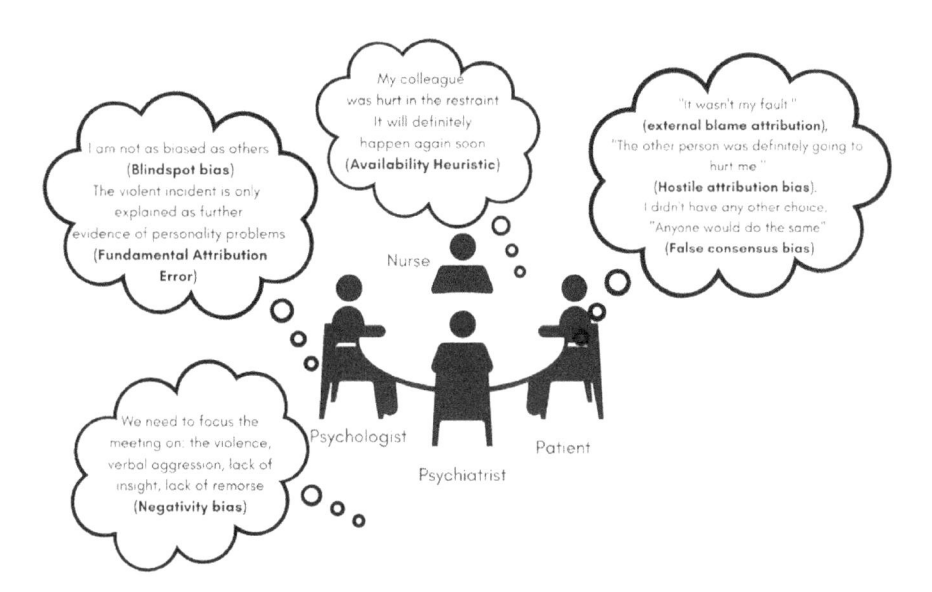

Figure 2.1 Multidisciplinary team meeting: examples of possible biases
N.B. The biases attributed to each role are for illustration only – these biases are applicable to us all, and these thoughts will not always be a bias.

behaviour that one is observing, but external causes (e.g. situation) to the behaviour that one is acting out. An observer might attribute Antony's rude behaviour to his personality characteristics. However, if they considered his situation from his perspective, they might instead make situational attributions when trying to understand how they would behave in a similar context. Putting team members in Antony's shoes can help to challenge correspondence bias as it encourages the consideration of alternative factors, besides his personality, to explain his behaviour.

An additional consideration in Antony's multidisciplinary team might be related to culture. For example, there is evidence that children in individualistic and collectivist cultures differ in internal and external attributions of behaviour (Miller, 1984). Western children tend to make external attributions about behaviour until late childhood, after which point they increasingly make internal attributions. On the other hand, South Asian children tend to move increasingly towards making external attributions about behaviour (ibid.). Therefore, if Antony's multidisciplinary team is comprised of people who were raised in a Western, individualistic culture, this may explain their correspondence bias in associating his behaviour to personality. This is just one example of how our experiences can influence the attributions that we make about events and behaviour, and shows that resulting biases are unique to our individual experiences.

How we Form Impressions of Other People

Our individual experiences shape the way in which we view and make sense of the world around us. We behave in ways that secure the success of achieving our objectives, including those which are reinforced and become automatic responses. Through our experiences, we learn how others behave to ensure the success of their own objectives, and make inferences about how groups of people are likely to behave to achieve the same outcome. In other words, we try to predict behaviour through making attributions. Making attributions at a group level is a cognitive shortcut to making sense of others' behaviour. While this can be effective when accurate, it can be unhelpful to those in, and associated with, the group. This group-level definition of people and the way they behave is also known as stereotyping (Hogg & Vaughan, 2018).

The serious problems associated with stereotyping are reflected in the disproportionate representation of individuals from different ethnic backgrounds in the criminal justice system (Goodman & Ruggiero, 2008), as well as in the overdiagnosis of schizophrenia among Black British, Black African, and Black African Americans, which was brought to light in the 1980s. It is well documented that a lack of understanding of cultural differences between the assessor and assessee, as well as a diagnostic classificatory system which was developed on the basis of White presentation, contributed to the medicalisation of Black people (e.g. Jones & Gray, 1986; Trierweiler et al., 2006). There is also evidence of implicit bias in the stereotypes that mental health practitioners hold when working with people across ethnic and cultural groups (e.g. Keating & Robertson, 2004; Lewis et al., 1990; Trierweiler et al., 2006). The stereotypes that we hold can be implicit (unconscious) or explicit (held in our awareness), and apply to different group memberships (e.g. race, gender, or age). This means that we can have an 'unconscious' stereotype about certain groups but hold a conscious awareness in contrast to this stereotype (Hogg & Vaughan, 2018). Therefore, the attributions that we make at a group level can sit in our subconscious, and influence how we make sense of events and behaviour and, ultimately, the decisions that we make. If we hold implicit biases towards groups, how do we then become consciously aware that they exist?

Several studies (summarised in Chapman et al., 2013) have used the Implicit Association Test (IAT) to explore bias among medical professionals. Overall, the findings point to conclusive evidence of a pro-White implicit bias among medical students (White-Means et al., 2009) and physicians (Sabin et al., 2008, 2009), which has a negative impact on clinical decision-making (Green et al., 2007) and patient satisfaction (Penner et al., 2010). Medical participants did not explicitly report a bias in favour of White patients, but their unconscious attributions still had a negative impact on practice. Findings of implicit bias among medical professionals can also occur in relation to gender, with some holding implicit pro-male biases in clinical decision-making (e.g. Borkhoff et al., 2008; Chapman et al., 2001).

Research on implicit bias among medical professionals has clear implications for practitioners within forensic services. Despite the competency, honesty, and hardworking nature of forensic clinicians involved in any stage of the CJS, their ability to work impartially, like that of those in any profession, is constantly under threat from their implicit biases. A plethora of research indicates that police hold pervasive implicit biases which link Black individuals to traits of aggression, violence, and hostility, and hold a type of attentional bias that sees them identify crime-related objects more quickly after subliminally perceiving black faces (Eberhardt et al., 2004). This, in turn, can increase police tolerance for violence against black individuals (Goff et al., 2014), and lead to disproportionate levels of stops, searches, and arrests in comparison to their White peers. This type of unconscious racial stereotyping is also prevalent among White healthcare providers who have been shown to associate White people with intelligence, success, and education, and Black people with aggression, impulsivity, and laziness (Wittenbrink et al., 1997). Moreover, Rachlinski et al. (2009) explored unconscious racial bias at the stage of conviction and sentencing. They put practicing judges through a series of tests, questionnaires, and case vignettes to explore whether race would impact on sentencing. The results showed that when White judges were not told of the defendant's race, but subtly primed with cues that implied that the defendant was Black, they appointed significantly harsher sentences. However, when White judges were explicitly told that the defendant *was* Black, and they thought the research was focusing on racial bias (information gleaned through post-discussions) they were more motivated to practice fairly, meaning the same discrepancies did not appear (Rachlinski et al., 2009). Collectively, this research paints a fairly dejected picture: not only do Black individuals disproportionately suffer at the mercy of authority's implicit biases, but those same authorities are likely unaware of the prejudice and harm they are enforcing.

Practitioners can also hold biases in relation to gender. The vignette below, in Context 3, illustrates how gender might influence individual and organisational responses to clients.

> The psychologist working with Antony also has a female patient on their caseload with the same diagnoses, similar background, and serious index offences. They have noticed that the prioritisation of treatment and framing for the two cases are quite different, with the female patient being recommended for trauma therapy, and the male patient for violence/offence-focused work. The psychologist starts to question in the clinical team whether there are organisational and cultural gender stereotypes which impact on treatment provision (i.e. intergroup attribution- women offend because they are traumatised, men offend because they are aggressive) and notes that there is much more staff training in trauma in the female service, with commissioners and review tribunals asking for different work for different genders.
>
> (Context 3)

As alluded to above, the influence of gendered cultures on an understanding of presenting patient need and provision of interventions challenges the care offered to Antony. Organisational cultures that provide safe spaces for challenge to attributions and biases are important in providing equitable support for people in forensic services (see Chapter 10).

Decision-Making in Forensic Settings

Decision-making plays a central role within forensic settings. This is particularly evident when professionals conduct assessments that directly inform the legal process. Pre-trial, this can involve assessing risk and personality/cognitive functioning to determine an individual's fitness to stand trial, or the setting to which they will be sentenced or directed (i.e. prison or a secure hospital). Thereafter, assessment and decision-making informs the level of security needed, whether escorted or unescorted leave can be granted, and, ultimately, when release is recommended – all of which is dependent on the type of sentence or direction the individual is under. In addition to risk assessment, forensic practitioners are continually involved in decision-making around formulating the treatment needs and intervention plans for individuals within the CJS.

Court systems utilise the findings and opinions of forensic evaluators, with over 95 per cent concordance rates between expert witnesses and subsequent legal decisions (Zapf et al., 2004). With such weight being given to expert testimony, and the high stakes involved in balancing public protection and restriction of liberty, it is important to understand and consider the role of bias in forensic decision-making (Zappala et al., 2018). While human bias has been studied for decades within the field of social psychology, researchers have argued that comparatively less research has been conducted and applied to forensic psychology (Beltrani et al., 2018). We will now turn to exploring some of the literature that helps us to understand attributions and biases in forensic contexts, and end with outlining a set of recommendations for how to protect against them.

Adversarial Affiliation

As already highlighted, bias can be both implicit (i.e. outside of a forensic evaluators' awareness) and explicit (i.e. within conscious awareness). Ideally, to ensure robustness in high-stakes decision-making, forensic assessments would have the same, or very similar, outcomes regardless of who conducted them. However, by virtue of being human, bias applies to us all- even qualified and highly trained forensic psychologists (Neal & Grisso, 2014).

Various studies have explored consistency between professionals who have assessed the same individual involved in the forensic legal process (e.g. Zapf et al., 2018) and found that it is not uncommon for assessors to not correspond. For example, Gowensmith et al. (2013) conducted research in Hawaii where the legal system dictates that defendants are assessed by three independent assessors to determine competency status (fitness to stand trial). Upon reviewing 216 such cases, they found that assessors did not agree on the competency status in 29 per cent of the cases. There was also no consensus among 45 per cent of 165 insanity cases. This research highlights that there is variation in the conclusion of forensic assessments for the same individual. Whether this difference is a result of the reliability of the assessment tools being used, or possible bias, it remains an important matter to unpick (Zapf et al., 2018).

One area of research that helps us to do this is linked to adversarial affiliation (Gowensmith & McCallum, 2019; Kamorowski et al., 2021; Murrie & Balusek, 2008) – the idea that we may be influenced by those who commission an assessment (e.g. a legal representative or a detaining authority). It seems fair to wonder if professionals employed by a legal team who work on behalf of a prisoner are more likely to rate that prisoner's risk as lower than a professional who is independent from the legal team. Similarly, are professionals who are employed by a prison or secure hospital more likely to rate risk levels higher than a professional independent from the detaining authority?

There have been both field and experimental studies that have aimed to explore whether such influence exists, or whether forensic evaluators are able to remain objective regardless of the commissioning authority. Research examining inter-rater reliability of risk assessment tools has broadly shown that high levels of agreement can be achieved by two or more assessors

(Barbaree et al., 2001; Doren, 2004, 2006; Hanson, 2001). Interestingly, in the context of forensic-clinical practice, several studies have shown that differences in inter-rater scores start to emerge when inter-rater agreement is explored in adversarial legal proceedings. For example, through exploring the use of the Psychopathy Check List–Revised (PCL-R; a measure used to explore the personality construct of psychopathy, often as part of a risk assessment) in 23 cases of sexual violence in the United States, where two evaluators (from opposing sides) had assessed an individual, Murrie et al. (2008) found that the level of agreement was 0.39; significantly below the agreement achieved in research settings (roughly 0.85). The difference in scores was predominantly in line with the commissioning 'side'; that is, the prosecution-instructed assessor scored higher than the defence-instructed assessor. These results were further expanded upon by Murrie et al. (2009) who examined the application of various psychological assessment/risk assessment tools (e.g. STATIC-99, PCL-R, and MnSOST-R) in cases of sexual violence, and found that inter-rater agreement was lower than that reported in research-context studies, and thus supported evidence of adversarial allegiance.

This research is incredibly important to reflect on. There should be an attempt to understand this 'pull' to the commissioning side to ensure that forensic practitioners can mitigate against unconscious bias (Murrie & Balusek, 2008). Perhaps, first, by asking why there is a difference in assessment outcomes depending on who commissions a report, and what processes may be at play here. To do this, let us turn back to Antony and consider Context 4 where he faces a Mental Health Review Tribunal for which he was assessed by two different affiliations:

> Antony is having a Mental Health Review Tribunal. He has been assessed by a Forensic psychologist from the hospital (detaining authority) and by a solicitor-instructed psychologist. The psychologists have assessed Antony using the same structured professional judgement tool (the HCR-20) but have both arrived at different risk levels and recommendations. The psychologist for the detaining authority has rated Antony's risk as higher than the solicitor-instructed psychologist, and has not recommended discharge, unlike the other assessor. The psychologist from the detaining authority (secure hospital) thinks that the solicitor-instructed psychologist is biased because they are being paid to support the case for discharge. They have not considered that they too may hold biases in the other direction.
>
> (Context 4)

It is likely that there are two forms of cognitive bias at play here. Firstly, as each assessor appears to have made an evaluation that favours their affiliation (i.e. higher-risk evaluation for a secure hospital, and lower-risk evaluation for the legal team advocating for Antony's discharge), it is probable that they are both demonstrating *adversarial affiliation bias* (Burke, 1966) – the tendency for expert opinion to be influenced in favour of the party who they represent. The second bias is described in the following section.

Blind Spot Bias

In Context 4, above, the secure hospital psychologist believes that the assessment made by their solicitor-instructed counterpart is biased, while simultaneously failing to see how their own judgement could be subject to a similar influence. Therefore, it is likely that they are practicing *blind spot bias*- the tendency to identify bias in others while failing to recognise those same biases in oneself (Pronin et al., 2002). Blind spot bias is a particularly common feature of psychological evaluations. For example, Zapf et al. (2018) surveyed 1099 mental health professionals from a total of 39 countries who had a role of conducting forensic mental health assessments for legal processes. Professionals were asked about their beliefs about the prevalence and nature of cognitive bias in forensic mental health evaluation, and it was found that 79% of them agreed that cognitive bias requires serious consideration for forensic mental health evaluation, but that only 52% of them believed that it could influence their own judgements.

Availability Heuristic

Another way in which bias can influence individual decision-making is through the availability heuristic (Hogg & Vaughan, 2018). An example of this bias is illustrated in the vignette below, which centres around decision-making for a critical juncture in Antony's care pathway:

> The clinical team for Antony is discussing whether they think there should be an application to the Ministry of Justice for escorted leave as he has been engaging well with treatment, and without incident, for some time. The whole team are supportive in principle of this next step. However, the Consultant Forensic Psychiatrist who is the Responsible Clinician (RC) for Antony has serious reservations. The RC had a difficult case three years ago. They made the decision to discharge a patient who had the same diagnosis as Antony, and went on to commit a further very serious act of violence. This case keeps coming to mind, and is making the decision about Antony very difficult. The RC thinks that Antony is still too high risk for escorted leave. The RC takes the case to a colleague for peer discussion, and is asked to reflect on how much the previous case is impacting their decision-making.
> (Context 5)

In Context 5, it seems that the RC's reservations about Antony are being driven by *the availability heuristic* – the tendency to use information that is most readily available to inform our judgements and decisions. In this case, the RC's previous experience of discharging a patient who went on to commit a serious offence (a false negative, incorrectly assessing low risk) seems to be having a direct impact on decision-making in relation to Antony. False negatives are likely to be more memorable than false positives (e.g. incorrectly assessing high risk) because if an inaccurate assessment of low risk of reoffending is made, and then that person goes on to offend, this is likely to remain with a clinician in a way that it probably would not for somebody who they assessed as likely to reoffend but did not. This then poses the risk of the clinician overestimating future risk in their assessments.

A more rudimentary example of the availability bias, as illustrated by Ruscio (2000), is when most people believe that the chances of dying from a shark attack are greater than the likelihood of dying from falling aeroplane parts, despite the contrary being true. However, because the former receives more media attention than the latter, the idea of dying from a shark attack is likely to come to mind faster (the bias is more available) than dying by falling plane parts, and inform, what is in fact, an incorrect judgement.

Representativeness Heuristic

Context 5 also illustrates an example of the representativeness heuristic. The representativeness heuristic is a cognitive shortcut which entails calculating the likelihood of an event based on its similarity to previous events (Tversky & Kahneman, 1974). When faced with uncertainty, which is often the case when working to understand and manage complex forensic presentations, we have a tendency to focus on similar features to our past experience, and draw conclusions based on those similarities. In Context 5, the RC could be influenced by the representativeness heuristic, and thus be placing an over-emphasis, albeit unconsciously, on features of the previous case that are alike to Antony's violent offending, and diagnosis of schizophrenia. When potential sources of bias are brought to consciousness, this allows for full and robust exploration of features which differentiate the individual circumstances.

Framing and Anchoring Effects

Framing is a cognitive heuristic which highlights how people form opinions based on the initial presentation of a situation (Hogg & Vaughan, 2018). People may make different interpretations on the basis of the same information, depending on how the information was framed and delivered. You can see how this links to the adversarial affiliation and related biases (discussed in an earlier section of this Chapter), as there is a strong organisational and/or legal context surrounding the evaluator. Furthermore, we tend to be most strongly influenced by the information that we first encounter (ever heard phrases relating to the importance of 'first impressions'?), to the point that subsequent information that might be contradictory to this is sometimes overlooked. This cognitive phenomenon is referred to as the anchoring effect; it is difficult for people to adjust their initial positions even in the face of new evidence (Hogg & Vaughan, 2018). Context 6 sets out how framing and anchoring effects might influence decision-making in forensic practice:

> A psychologist has been tasked with completing an assessment of risk for Antony. Antony has not been violent for three years. He has had one verbal altercation with a fellow patient two and a half years ago where he threatened violence. This incident is written in every report, explained in great detail, and becomes a significant focus of the risk assessment (availability heuristic). The referral for the risk assessment from the psychiatrist was presented as examining progress of a 'high risk violent offender with a significant history of violent offending, and current offence paralleling behaviour'. This unconsciously creates an initial lens through which the psychologist views and approaches the assessment (anchoring effect and framing).
>
> (Context 6)

Confirmation Bias

Confirmation bias is the tendency to seek evidence that fits with our view of the world (MacLean et al., 2019). In forensic practice, we must be cautious about seeking evidence that confirms ideas that we have about the clients with whom we work. This is illustrated in Context 7 below:

> You start to monitor your risk assessments, and realise that you have rated Antony as the same risk of violence (high) as another patient. By consciously reviewing your own decision-making, you notice some interesting patterns. Antony has not been violent for two years. The other patient continues to engage in violence every few months. You notice that the risk assessment documents state that Antony is argumentative, unmotivated, and unappreciative of staff. The other patient however is open, apologetic (after an incident), and expresses gratitude towards staff for their care. You start to question whether the difference in presentation is truly reflective of risk or whether you are being unconsciously biased by how you feel in response to the different presentations. Is there confirmation bias where you look for the negative with Antony and positive for the other patient?
>
> (Context 7)

In Context 7, there could be bias to interpret behaviours based on the way in which each client is viewed, and the response that this triggers in the professional. There is one patient that has regular violent incidents but who is accessing and valuing care. For a caring professional, this is likely to enhance self-esteem and feelings of clinical competence. On the other hand,

Context 7 presents a client who has not displayed regular violence, but who is unappreciative and difficult to form a connection with. When assessing risk, it is possible that our judgements could be impacted by confirmation bias; where we look for evidence, albeit unconsciously, to support the view that we have made. Someone may be viewed as 'riskier' because they are harder to connect with, or less risky because they are instantly likeable, which, in reality, may or may not *actually* relate to level of risk. Clinicians may draw conclusions based on underdeveloped hypotheses, and may not spend the necessary time needed to robustly test their hypotheses.

Hindsight Bias

Hindsight bias relates to our tendency to use the outcome of an event to form a view of how predictable the event would have been with foresight alone (Hoff & Vaughan, 2018). In exploring this type of bias, LeBourgeois et al. (2007) provided 235 general and forensic psychiatrists with case information, and asked them to rate risk. One group had outcome information (i.e. they were told that the patient engaged in an adverse behaviour) and the other group were not given an outcome. Those who were told that the patient engaged in risk behaviour rated risk significantly higher than those who were not. This study found that hindsight bias was more prevalent in general psychiatrists compared to forensic psychiatrists. Upon further examination by Beltrani et al. (2018), who aimed to replicate these findings, 95 forensic mental health evaluators were divided into two groups – the hindsight group had an additional sentence regarding the outcome of the case which was not present for the foresight group. It was found that participants with outcome information made risk ratings that were significantly higher. Additionally, this group were more likely to suggest that they would have pre-empted violence, compared to those without the outcome information. Context 8 sets out how this might apply to Antony:

> Antony progressed to unescorted leave after a risk review following on from engaging in therapy, and having had a long period of time without any violence. One day on the ward a patient (who was very unwell) became verbally aggressive towards Antony, and threatened to kill him. Antony then punched the patient several times. During the ward round, the risk scenarios and levels were reviewed on the risk assessment (HCR-20) and several members of the team questioned how this scenario had not been fully planned for because, to them, it was 'obvious' that this would happen, and that it was 'only a matter of time'.
> (Context 8).

Decision-making is complex, and even more so when it involves assessing the likelihood of certain human behaviours. It is imperative that psychologists and other professionals in the CJS monitor their perception of foresight and hindsight. Overestimating our ability to predict outcomes, and underestimating the complexity of the process, may mean we think we are better at risk 'prediction' than we really are. The scenario emphasises the natural tendency for professionals to reassess what we 'should' have known with new information that was not available at the time.

Summary

This list of attribution biases is ever-expanding, and we have not been able to explore them all here. For example, it is common to observe the illusory correlation (i.e. seeing a relationship between two variables when no relationship exists), false consensus effect (i.e. seeing our own behaviour as more typical than it really is), self-serving bias (i.e. seeing our strengths as internal, and difficulties as external, and vice versa for others), intergroup attribution (i.e. assigning the cause of one's own or others' behaviour to group membership), and hostile attribution bias (i.e. being more likely to assign people's behaviours/motivations as hostile in intent, rather than neutral or unharmful; Försterling, 2001; Hogg & Vaughan, 2018; Stainton Rogers, 2020). Figure 2.1 illustrates what some of the different types of biases may look like in a multidisciplinary setting.

Understanding Bias Mitigation

Entrenched in the architecture of one's mind from years and years of social experience, bias is a difficult tendency to mitigate (Anderson, 2010). Unconscious bias, in particular, can be thought of as a habit of the mind, strongly resistant to change. As has been shown throughout this Chapter, there are a variety of biases that can impact forensic practice- each capable of being motivated by a range of different factors (Deitchman et al., 1991; Miller et al., 2011; Murrie et al., 2008). However, it is important to realise that bias will always exist in human decision-making. The focus, therefore, should not be on trying to eradicate it from human cognition, but on trying to understand it, the threat it poses, and what sensible strategies can be developed to address it. This should be considered as a necessary tenet of good forensic practice.

The need here to refer to the evidence in mitigating bias is important to highlight. Although forensic practitioners are aware of explicit bias, they perceive their own judgements to be less prone to bias, and underestimate the ability for bias to influence judgements (Neal & Brodsky, 2016; Kukucka et al., 2017). In addition to this, practitioners seem to underestimate the efficacy of evidence-based bias-reduction techniques. When forensic evaluators were asked about how they combat their bias (Neal & Brodsky, 2016), the primary strategy identified was *'thinking about bias'* (i.e. introspection), despite evidence showing this to be ineffective (Pronin & Kugler, 2007). Similarly, in a survey of 351 forensic psychologists regarding strategies for combating bias in forensic evaluation, 100 per cent of them reported attempting to minimise bias through introspection. In support of these results, Zapf et al. (2018) found that 87 per cent of professionals believed that they could minimise bias by simply trying to set it aside. Pronin and Kugler (2007), however, argue that this is *introspection illusion*- the belief that one can combat bias by merely thinking about doing so. The irony here, however, is evident; if, by definition, implicit bias does not exist at a conscious level, then one cannot plausibly, and consciously, think about combating it. Altogether, this research indicates that the ability for evaluators to understand and adequately monitor their own potential for bias in forensic work is extremely limited.

Recommendations for Mitigation in Forensic Practice

Neal and colleagues set out to conduct a meta-analysis exploring literature of bias and de-biasing techniques in forensic mental health practice, only to conclude that the evidence pool was not wide enough to allow for this, and so instead engaged in a systematic review (Neal et al., 2022). This highlights that further research is needed to understand effective de-biasing techniques that are specific to forensic practice (ibid.).

So, what might work to debias forensic practice based on what we do know? Drawing on a range of studies we attempt to answer this question by outlining the following suggestions:

1 Know our own biases.
2 Training and supervision.
3 Monitoring logs.
4 Information gathering and communication.
5 Hypothesis testing.
6 Evidence-based tools.
7 Base rates.
8 Multiple assessors.

Know our Own Biases

The first debiasing step should be for professionals to actively become aware of the implicit stereotypes that they may hold (Burgess et al., 2007; Kamorowski, et al., 2021; Zapf & Dror, 2017). This could involve participation in measures such as the Implicit Associations Test (IAT), which can help individuals to identify any bias tendencies, and form motivation to address them. Understanding the lens through which *we* see the world is crucial. We can do this through reflecting on how our life has impacted the way we see the world, including our experiences, attachments, gender, race, and social and cultural background. In turn this can help us as professionals to consider the way in which we approach interventions and assessments in forensic settings. From an early stage of training, we need to encourage familiarity

Figure 2.2 De-biasing techniques

with our own filters, and understand how these will inevitably interact with the filters of others around us. Supervision and reflective practice should actively encourage practitioners to share and support one another to explore how biases could be operating in practice.

It is important that a compassionate approach is taken during this process to encourage people to understand the nature of bias in a way that helps them feel able to openly acknowledge their biases without feeling threatened. Stereotyping and partial judgements are a normal part of human cognition. From an evolutionary perspective, we have inherited a brain that processes and works in a way that is outside of our design and control (Gilbert, 2014). Erroneous thinking happens to us all. The focus should be on understanding how we can reduce the harm associated with bias; a focus that, in turn, functions to reinforce the integrity of forensic practice.

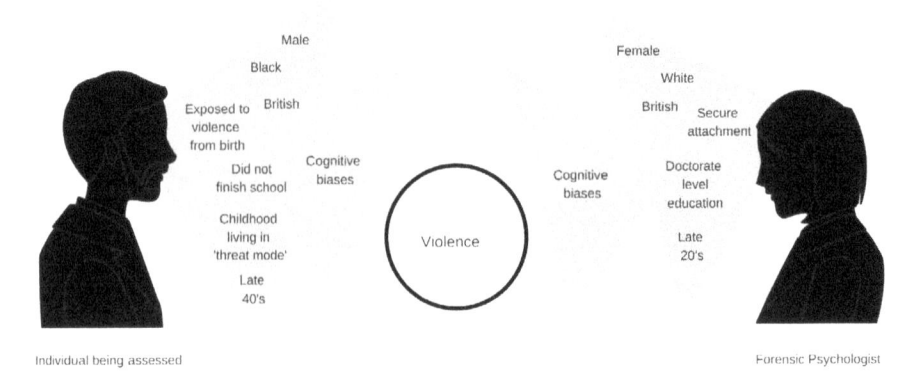

Figure 2.3 An illustration of how our 'lenses' may filter our view of the same topic and how our experiences shape our biases

Training and Supervision

Understanding our biases is important but will not address the complexity of such cognitive shortcuts alone. Another important way to mitigate against bias is through training. Such training includes (but is not limited to):

- **Understanding bias**: Training in bias, how it operates, and the factors that develop effective debiasing strategies should arguably be a central part of training to becoming a forensic practitioner (Dror, 2009). Supervision should provide the ongoing space for reflection.
- **Risk assessment/induction training**: Such training should aim to educate individuals about the normality of making mental shortcuts, and their susceptibility to implicit biases. A focus on heuristics should be included. It should also help individuals to understand *how* to make these biases conscious and actively address them as part of risk assessment, and how to reflect this in the report writing process.
- **Cultural competence**: In addition to training in understanding bias, there should be a concerted effort to train practitioners to be culturally competent (Kapoor et al., 2013). Forensic practice has further to go in terms of ensuring that there is representation of the population served, and recognising the role of culture is crucial. This should also form a core part of supervision skills training.

Monitoring Logs

Those working in forensic practice should monitor and record their decision-making to become aware of their patterns (Gowensmith & McCallum, 2019). Are you someone who errs on the side of caution, and rarely makes recommendations for progressive moves? Are there particular offences/

personality presentations that influence your decision-making? Keeping a record of recommendations made for risk assessments (including who the commissioning authority was) or recording the decision-making process as part of a multidisciplinary team for leave/moving to lesser security, can help professionals develop their baseline (i.e. to understand their key common tendencies, such as when they recommend progression and when they do not). This awareness then creates the opportunity for noticing patterns, understanding the factors that may influence such patterns, and taking active steps to ensure that there are robust strategies to 'test' the objectivity of decision-making.

Information Gathering and Communication

At the stages of pre-evaluation and evaluation, it is crucial that forensic evaluators seek a variety of information from a range of sources to ensure the robustness of their eventual assessment. We form initial impressions quickly which means that there can be a pull towards confirmatory bias where evidence is sought to support, and not challenge, an initial formulation. This increases one's chances of ignoring salient information which can result in a one-sided, misinformed, and discreditable evaluation. It is therefore important to vigorously seek evidence that may both *con*firm and *dis*confirm hypotheses (Borum et al., 1993). It may also be helpful to record one's impressions of each new piece of information to monitor thought processes. Be aware, however, that the information we review first tends to be more powerful than subsequent information (i.e. the anchoring effect).

Close attention should also be paid to the manner in which results are communicated. A study by Slovic et al. (2000) found that decision-makers perceived higher risk when an evaluator reported risk estimates in terms of frequency (e.g. '4 out of 10') rather than probability (e.g. '40%'), despite being equivalent. Becoming aware of *how* results are communicated and understood enhances insight into the role that bias may play in practice.

Furthermore, a study by Martin and colleagues explored the written notes of forensic mental health nurses in a secure hospital in Canada to look for qualitative evidence of bias in the way they record the care and treatment of patients (Martin et al., 2020). This study found evidence of negative biases in the notes whereby nurses regularly used language that implied they disbelieved patient experiences (discounting), and pathologised behaviours (fundamental attribution). This is not only interesting in terms of the role of nurses, but also due to fact that bias can cascade to other professionals (ibid.). This also links with other biases (e.g. framing) that have already been outlined. If a psychiatrist first reads about the behaviour of a patient from the nurses' notes, and the nurses' notes have interpreted behaviour as untruthful, this will, without active challenging, frame the way in which the psychiatrist makes sense of the event. Interesting ideas have been suggested by Martin et al. to counteract such bias in record-keeping, including having an automated prompt to review entries for personal interpretation and bias.

Hypothesis Testing

It is important to ensure that, as forensic practitioners, we revert to the scientist-practitioner model to develop and test clear hypotheses (Neal & Grisso, 2014). By consciously developing and recording hypotheses linked to behaviours, there can be plans for how to test them in a way that works towards the goal of understanding risk/distress/enhancing protective factors. Another aim is to actively 'consider the opposite'. This means that practitioners should ensure that they identify alternative hypotheses, and test them as robustly as they have tested their own (Dror, 2009).

This suggestion is supported by Griffith (2019) who aimed to assess the presence of confirmation bias (i.e. whether clinicians seek evidence to support their initial opinion formed). Once forensic mental health clinicians were presented with a case study they were asked to form an initial hypothesis, rate their confidence of this hypothesis, and then select the clinical information that they would like to see next from a list. The next stage of the study was to offer a short 'consider the opposite' intervention, and then re-rate the information they would like to see. Finally, clinicians were asked to offer a second hypothesis and confidence rating. The study showed a presence of confirmation bias (clinicians initially sought information that would confirm their hypothesis), and showed the de-biasing technique of 'consider the opposite' to be effective (clinicians did not demonstrate confirmation bias in the way they sought evidence post-intervention). This is promising as the de-biasing technique was brief (participants were asked to hold an opposite hypothesis and consider three reasons why this hypothesis could be accurate) and is therefore generalisable (Griffith, 2019).

Evidence-Based Tools

When conducting psychological assessments for risk/need/clinical presentation, it is crucial to use evidenced-based tools, and record how you have acknowledged and addressed bias. Although such tools do not free us from bias (as illustrated in links to literature throughout this Chapter), it is recommended that the user manual is reviewed *every* time to ensure no 'pull to affiliate' (i.e. gradually rewriting definitions of criteria and clinical opinions over time in favour of the party that employed you; Zapf & Dror, 2017). The active process of acknowledging and addressing the issue of bias will help to form a more robust report, and encourage practitioners to be conscious of the human processes that occur.

In addition to this, it is important to constantly remind ourselves that we are more likely to attribute the way someone behaves in a situation to factors relating to the person, rather than the environmental context (Hogg & Vaughan, 2018) This fundamental attribution error is not a reflection of our skills as professionals, but of a process that we are all susceptible to as human beings. As such, we need to do more to actively monitor this as part of the assessment and treatment process. When conducting risk assessments, for example, we could ensure adequate time is spent assessing and formulating contextual elements of any offending behaviour (e.g. social and situational factors).

Base Rates

A strategy that has already proved important is the encouragement of clinicians to focus on base rates- considering evidence from statistics, rather than just professional judgement. In clinical contexts, experts often underutilise or disregard base rate information and thus place more emphasis on case specific information (Neal & Grisso, 2014). Therefore, there should be an aim to make a conscious effort to form a clinical impression, compare it with base rate statistics, and challenge the information that creates the difference between the two.

Multiple Assessors

For particularly contentious/high stake/high profile cases, it may be helpful to have multiple assessors involved. This often happens in an adversarial manner, such as prosecution vs. defence, or detaining authority/prison versus legally instructed psychologists, but nonetheless comes with its own inherent biases (e.g. adversarial affiliation). It could be recommended that, as part of the process of assessment, two psychologists review the available information, and then reflect on the decision-making process together, making it different from the supervisory process. This is, however, resource-intensive and would certainly not be practicable in all cases.

An extension of this approach is 'blinding' techniques, whereby assessors/decision-makers are not exposed to information that could be task-irrelevant and biasing (Neal et al., 2022). For example, having file information reviewed by another person may help to ensure that it only includes relevant case material, and thus minimising the impact of irrelevant material on decision-making. Notably, however, this approach has practical, resource, and legal limitations (ibid.). A forensic practitioner may feel it important as the assessor that they are able to make judgments about what information they see/use as part of an assessment/to inform treatment. A middle ground to address this issue, while still mitigating bias, is outlined by Neal et al. (ibid.). Linear sequential unmasking is a technique in which the clinician will receive all information, but such information will be sequenced by another professional to minimise bias (Dror et al., 2015). For example, a psychologist conducting a risk assessment may receive information from the commissioning source in a staged manner. This would enable the assessor to record how new information impacts on their decision as they receive it. Another example could relate to a psychologist deciding to interview a patient/prisoner without reading previous assessments by other professionals (Neal et al., 2022). The psychologist may only receive any immediate risk information that is necessary for their safety during the interview process, and then be free to form their own initial opinion before starting to review other information.

In essence, there are a number of strategies that are important to consider, and as a profession we need to reflect on whether we are currently doing enough to understand and mitigate against bias.

Summary

This chapter has provided an overview of attribution and bias and its relevance to forensic practice. It has highlighted that, as forensic practitioners, we are not free from bias, and, given our roles in making high stakes decisions that impact on individual's liberty and the protection of the public, it is crucial that we develop our understanding and take reflective action.

In line with understanding the brain and natural human shortcuts we take, it is also crucial that forensic practitioners embed this knowledge in the context of trauma-informed care (Jones & Willmot, 2002). In order to provide fair and objective assessment and treatment for individuals within the CJS, we need to acknowledge and actively address our biases. Our background, culture, gender, and experiences, all impact on the way we develop as practitioners, just as it does for the individuals whom we work with within the CJS. While more research into understanding the presence and nature of bias in forensic decision-making and on effective strategies to mitigate risk is needed, there are techniques we can, and *should*, adopt. By becoming familiar with bias and our own natural tendencies, we become open to actively monitoring and challenging our approach to assessment, formulation, and interventions.

While there is clearly further research needed to develop evidenced-based debiasing techniques that are resource-effective, we hope that the suggestions outlined here provide ideas, and start a process of reflection, discussion, and further change.

References

Anderson, E. (2010). *The imperative of integration*. Princeton University Press.

Barbaree, H., Seto, M., Langton, C., & Peacock, E. (2001). Evaluating the predictive accuracy of six risk assessment instruments for adult sex offenders. *Criminal Justice and Behavior*, 28, 490–521.

Beltrani, A., Reed, A. L., Zapf, P. A., & Otto, R. K. (2018). Is hindsight really 20/20? The impact of outcome information on the decision-making process. *International Journal of Forensic Mental Health*, 17(3), 285–296.

Borkhoff, C. M., Hawker, G. A., Kreder, H. J., Glazier, R. H., Mahomed, N. N., & Wright, J. G. (2008). The effect of patients' sex on physicians' recommendations for total knee arthroplasty. *Canadian Medical Association Journal*, 178(6), 681–687.

Borum, R., Otto, R., & Golding, S. (1993). Improving clinical judgment and decision making in forensic evaluation. *Journal of Psychiatry & Law*, 21, 35–76.

Burgess, D., van Ryn, M., Dovidio, J., & Saha, S. (2007). Reducing racial bias among health care providers: Lessons from social-cognitive psychology. *Journal of General Internal Medicine: JGIM*, 22(6), 882–887.

Burke, K. (1966). *Language as symbolic action*. University of California Press.

Chapman, E. N., Kaatz, A., & Carnes, M. (2013). Physicians and Implicit Bias: How Doctors May Unwittingly Perpetuate Health Care Disparities. *Journal of General Internal Medicine: JGIM*, 28(11), 1504–1510.

Chapman, K. R., Tashkin, D. P., & Pye, D. J. (2001). Gender bias in the diagnosis of COPD. *Chest*, 119(6), 1691–1695.

Cushman, F. (2019). Rationalization is rational. *The Behavioral and Brain Sciences*, 43, E28.

Deitchman, M. A., Kennedy, W. A., & Beckham, J. C. (1991). Self-selection factors in the participation of mental health professionals in competency for execution evaluations. *Law and Human Behavior*, 15(3), 287–303.

Doren, D. (2004). Stability of the interpretative risk percentages for the RRASOR and Static-99. *Sexual Abuse: A Journal of Research and Treatment*, 16, 25–36.

Doren, D. M. (2006). Inaccurate arguments in sex offender civil commitment proceedings. In A. Schlank (ed.), *The sexual predator: Law and public policy-clinical practice* (3rd edition). Civic Research Institute.

Dror, I. E. (2009). How can Francis Bacon help forensic science? The four idols of human biases. *Jurimetrics*, 50, 93–110.

Dror, I. E., Thompson, W. C., Meissner, C. A., Kornfield, I., Krane, D., Saks, M., & Risinger, M. (2015). Letter to the editor – context management toolbox: A linear sequential unmasking (LSU) approach for minimizing cognitive bias in forensic decision making. *Journal of Forensic Sciences*, 60(4), 1111–1112.

Eberhardt, J., Goff, P., Purdie, V., & Davies, P. (2004). Seeing Black. *Journal of Personality and Social Psychology*, 87(6), 876–893.

Festinger, L. (1957). *A theory of cognitive dissonance*. Stanford University Press.

Försterling, F., (2001). *Attribution: An introduction to theories, research, and applications*. Psychology Press.

Gilbert, P. (2014). The origins and nature of compassion focused therapy. *British Journal of Clinical Psychology*, 53(1), 6–41.

Goff, P., Jackson, M., Di Leone, B., Culotta, C., & DiTomasso, N. (2014). The essence of innocence: Consequences of dehumanizing Black children. *Journal of Personality and Social Psychology*, 106(4), 526–545.

Goodman, A., & Ruggiero, V. (2008). Crime, punishment, and ethnic minorities in England and Wales. *Race/ethnicity: Multidisciplinary Global Contexts*, 2(1), 53–68.

Gowensmith, W. N., & McCallum, K. E. (2019). Mirror, mirror on the wall, who's the least biased of them all? Dangers and potential solutions regarding bias in forensic psychological evaluations. *South African Journal of Psychology*, 49(2), 165–176.

Gowensmith, W. N., Murrie, D. C., & Boccaccini, M. T. (2013). How reliable are forensic evaluations of legal sanity? *Law and Human Behaviour*, 37(2), 91–106.

Green, A. R., Carney, D. R., Pallin, D. J., Ngo, L. H., Raymond, K. L., Iezzoni, L. I., & Banaji, M. R. (2007). Implicit bias among physicians and its prediction of thrombolysis decisions for black and white patients. *Journal of General Internal Medicine*, 22(9), 1231–1238.

Griffith, R. L. (2019). Forensic confirmation bias: Is consider-the-opposite an effective debiasing strategy? Unpublished master's thesis, Washburn University.

Hanson, R. K. (2001). *Note on the reliability of STATIC-99 as used by the California Department of Mental Health evaluators*. Unpublished report. California Department of Mental Health.

Heider, F. (1958). *The psychology of interpersonal relations*. Wiley.

Hogg, M., & Vaughan, G. (2018). *Social psychology* (8th edition). Pearson Education.

Jones, B. E., & Gray, B. A. (1986). Problems in diagnosing schizophrenia and affective disorders among blacks. *Psychiatric Services*, 37(1), 61–65.

Jones, E. E., & Davis, K. E. (1965). From acts to dispositions: The attribution process in person perception. In L. Berkowitz (ed.), *Advances in experimental social psychology* (2nd edition, pp. 219–266). Academic Press.

Jones, E. E., & Nisbett, R. E. (1972). The actor and the observer: Divergent perceptions of the causes of behavior. In E. E. Jones, D. E. Kanouse, H. H. Kelley, R. E. Nisbett, S. Valins, & B. Weiner (eds), *Attribution: Perceiving the causes of behavior* (pp. 79–94). General Learning Press.

Jones, L., & Willmot, P. (2002). The future of trauma informed forensic practice. In P. Willmot & L. Jones (eds), *Trauma-informed forensic practice* (pp 413–425). Routledge.

Kamorowski, J., De Ruiter, C., Schreuder, M., Ask, K., & Jelicic, M. (2021). Forensic mental health practitioners' use of structured risk assessment instruments, views about bias in risk evaluations, and strategies to counteract it. *International Journal of Forensic Mental Health*, 21, 1–19.

Kapoor, R., Dike, C., Burns, C., Carvalho, V., & Griffith, E. (2013). Cultural competence in correctional mental health. *International Journal of Law and Psychiatry*, 36(3–4), 273–280.

Keating, F., & Robertson, D. (2004). Fear, black people and mental illness: a vicious circle? *Health & social care in the community*, 12(5), 439–447.

Kelley, H. H. (1967). Attribution theory in social psychology. In D. Levine (ed.), *Nebraska symposium on motivation* (pp. 192–238). University of Nebraska Press.

Kukucka J., Kassin, S. M., & Zapf, P. A., & Dror, I. E. (2017). Cognitive bias and blindness: A global survey of forensic science examiners. *Journal of Applied Research in Memory and Cognition*, 6(4), 452–459.

LeBourgeois, H. I., Pinals, D. A., Williams, V., & Appelbaum, P. S. (2007). Hindsight bias among psychiatrists. *Journal of the American Academy of Psychiatry and the Law*, 35, 67–73.

Lewis, G., Croft-Jeffreys, C., & Anthony, D. (1990). Are British psychiatrists racist? *The British Journal of Psychiatry*, 157(3), 410–415.

MacLean, N., Neal, T., Morgan, R., & Murrie, D. (2019). Forensic clinicians' understanding of bias. *Psychology, Public Policy, and Law*, 25(4), 323–330.

Martin, K., Ricciardelli, R., & Dror, I. (2020). How forensic mental health nurses' perspectives of their patients can bias healthcare: A qualitative review of nursing documentation. *Journal of Clinical Nursing*, 29(13–14), 2482–2494.

Miller, J. G. (1984). Culture and the development of everyday social explanation. *Journal of Personality and Social Psychology*, 46, 961–978.

Miller, A. K., Rufino, K. A., Boccaccini, M. T., Jackson, R. L., & Murrie, D. C. (2011). On individual differences in person perception: Raters' personality traits relate to their psychopathy checklist-revised scoring tendencies. *Assessment*, 18, 253–260.

Murrie, D. C., & Balusek, K. (2008). Forensic assessment of violence risk in adversarial proceedings: Pursuing objectivity and avoiding bias. *Journal of Forensic Psychology Practice*, 7(4), 141–153.

Murrie, D. C., Boccaccini, M. T., Johnson, J., & Janke, C. (2008). Does interrater (dis)agreement on Psychopathy Checklist scores in sexually violent predator trials suggest partisan allegiance in forensic evaluations? *Law and Human Behavior*, 32, 352–362.

Murrie, D., Boccaccini, M., Turner, D., Meeks, M., Woods, C., & Tussey, C. (2009). Rater (dis)agreement on risk assessment measures in sexually violent predator proceedings: Evidence of adversarial allegiance in forensic evaluation? *Psychology, Public Policy, and Law*, 15(1), 19–53.

Neal, T., & Brodsky, S. (2016). Forensic psychologists' perceptions of bias and potential correction strategies in forensic mental health evaluations. *Psychology, Public Policy, and Law*, 22(1), 58–76.

Neal, T., & Grisso, T. (2014). The cognitive underpinnings of bias in forensic mental health evaluations. *Psychology, Public Policy, and Law*, 20(2), 200–211.

Neal, T., Lienert, P., Denne, E., & Singh, J. P. (2022). A general model of cognitive bias in human judgment and systemic review specific to forensic mental health. *Law and Human Behaviour*, 46(2), 99–120.

Penner, L. A., Dovidio, J. F., West, T. V., Gaertner, S. L., Albrecht, T. L., Dailey, R. K., & Markova, T. (2010). Aversive racism and medical interactions with Black patients: A field study. *Journal of experimental social psychology*, 46(2), 436–440.

Pronin, E., Lin, D. Y., & Ross, L. (2002). The bias blind spot: Perceptions of bias in self versus others. *Personality and Social Psychology Bulletin*, 28, 369–381.

Pronin, E., & Kugler, M. B. (2007). Valuing thoughts, ignoring behavior: The introspection illusion as a source of the bias blind spot. *Journal of Experimental Social Psychology*, 43(4), 565–578.

Rachlinski, J., Johnson, S., Wistrich, A., & Guthrie, C. (2009). Does unconscious racial bias affect trial judges? *The Notre Dame Law Review*, 84(3), 1195–1246.

Ross, L., Greene, D., & House, P. (1977). The 'false consensus effect': An egocentric bias in social perception and attribution processes. *Journal of Experimental Social Psychology*, 13, 279–301.

Ruscio, J. (2000). The role of complex thought in clinical prediction. *Journal of Consulting and Clinical Psychology*, 68(1), 145–154.

Sabin, J. A., Nosek, D. B. A., Greenwald, D. A. G., & Rivara, D. F. P. (2009). Physicians' implicit and explicit attitudes about race by MD race, ethnicity, and gender. *Journal of Health Care for the Poor and Underserved*, 20(3), 896–913.

Sabin, J. A., Rivara, F. P., & Greenwald, A. G. (2008). Physician implicit attitudes and stereotypes about race and quality of medical care. *Medical Care*, 46(7), 678–685.

Slovic, P., Monahan, J. & MacGregor, D. G. (2000). Violence risk assessment and risk communication: The effects of using actual cases, providing instruction, and employing probability versus frequency formats. *Law and Human Behavior*, 24(3), 271–296.

Stainton Rogers, W. (2020). *Perspectives on social psychology: A psychology of human being*. Routledge.

Trierweiler, S., Neighbors, H., Munday, C., Thompson, E., Jackson, J., & Binion, V. (2006). Differences in patterns of symptom attribution in diagnosing schizophrenia between African American and non-African American clinicians. *American Journal of Orthopsychiatry*, 76(2), 154–160.

Tversky, A., & Kahneman, D. (1974). Judgment under uncertainty: Heuristics and biases. *Science*, 185, 1124–1131.

Weiner, B. (1986). *An attributional theory of motivation and emotion*. Springer.

White-Means, S., Dong, Z., Hufstader, M., & Brown, L. T. (2009). Cultural competency, race, and skin tone bias among pharmacy, nursing, and medical students: implications for addressing health disparities. *Medical Care Research and Review*, 66(4), 436–455.

Wittenbrink, B., Gist, P., & Hilton, J. (1997). Structural properties of stereotypic knowledge and their influences on the construal of social situations. *Journal of Personality and Social Psychology*, 72(3), 526–543.

Zapf, P. A., & Dror, I. E. (2017). Understanding and mitigating bias in forensic evaluation: Lessons from forensic science. *International Journal of Forensic Mental Health*, 16(3), 227–238.

Zapf, P. A., Hubbard, K. L., Cooper, V. G., Wheeles, M. C., & Ronan, K. A. (2004). Have the courts abdicated their responsibility for determination of competency to stand trial to clinicians? *Journal of Forensic Psychology Practice*, 4(1), 27–44.

Zapf, P. A., Kukucka, J., Kassin, S. M., & Dror, I. E. (2018). Cognitive bias in forensic mental health assessment: Evaluator beliefs about its nature and scope. *Psychology, Public Policy, and Law*, 24(1), 1–10.

Zappala, M., Reed, A. L., Beltrani, A., Zapf, P. A., & Otto, R. K. (2018). Anything you can do, I can do better: Bias awareness in forensic evaluators. *Journal of Forensic Psychology Research and Practice*, 18(1), 45–56.

3 Social Identity Theories

Deborah Morris and Elanor Webb

Introduction

Group membership arises from an evolutionary need to belong, self-protect, survive, and thrive (Over, 2016). From our *group* membership, it is argued that *individual* self-concept, identity, beliefs, motivations, and behaviour can be understood. The relationship between 'I' and 'we', and how group membership impacts on self-identity and behaviour, is of long-standing interest to social psychologists (Hogg, 2018). Group membership and its relationship to self-concept and offending behaviour is also a core concern for those working with forensic populations.

Group membership is broadly defined as a "set of individuals who hold a common social identification or view themselves as members of the same social category"[1] (Stets & Burke, 2000, p. 225). The criterion for social category is broad. It can range from temporary membership based on transient criteria and time-limited activities, through to more enduring, stable, or demographic membership variables. Group membership can also constitute the *perception* of a collective shared social identity (SI), which, in turn, can facilitate group behaviour (Hogg & Rinella, 2019). Social psychological experiments have consistently demonstrated that group membership has far-reaching consequences for self-concept, and for relationships and behaviour between and within groups across settings, including healthcare (Bochatay et al., 2019). While different theories have been developed to account for group behaviour and its relationship with self-concept, SI approaches remain the dominant 'meta-theory' (Hogg & Vaughn, 2018).

Social Identity Theories

From their inception, SI theories have been concerned with attempting to account for the extreme behaviours of *individuals* enacted in the context of *group* membership. As such, they are well positioned to account for behaviours that fall outside of social norms, including offending in the context of mental disorder. They are rooted in post-World War II political dogma that attempted to understand Nazism and the psychosocial processes that made

DOI: 10.4324/9781315560243-3

acts, such as genocide possible. In this context, initial models that focused on individual factors, such as the Authoritarian Personality (Adorno et al., 1950), were developed. They promoted the belief that behaviour in the Nazi regime was the manifestation of individual [personality] dysfunction- a belief that was consistent with post-war formulations that emphasised the 'Germans are different' hypothesis. Sherif's (1966) Realistic Conflict Theory adopted a more objective approach, prompting a shift towards a dynamic group-based understanding of behaviour.

As a European meta-theoretical model, SI theories stress that social context is key to understanding individual and group behaviour and offer a comprehensive analysis of the relationship between self and group, the function of group membership, and relationships within and between groups (Hogg, 2018). Central to SI theories is the assumption that cooperation, conflict, discrimination, and prejudice are group phenomena which are greater than the sum of individuals. Within SI approaches, the 'self' is defined as a social construct, arising from group [social] cognitive processes. SI approaches draw on two compatible models- the social identity theory, which focuses on intergroup relations and social change, and the self-categorisation theory, which focuses on cognitive processes- to account for intergroup processes (Hogg, 2018). These theories and their relevance to forensic practice will now be explored.

Social Identity Theory and Social Identity Theory of Intergroup Behaviour

Synthesising initial findings from group experiments, Social Identity Theory (SIT; Tajfel & Turner, 1979, 1986) was developed. SIT seeks to account for self-concept, motivations underlying group membership, intergroup behaviours (e.g. conflict and cooperation), and processes for social development (Haslam et al., 2012).

Central to SIT are the assumptions that society is structured by social groups and that group membership generates a SI (Hogg & Vaughn, 2018). SIT stresses that, though separate, SI and personal identity are experienced as an integrated whole, and are functionally interdependent (Turner & Reynolds, 2012). Group membership prescribes the beliefs and behaviours for individual members, which conform to the salient characteristics of the group. While the characteristics of groups differ, they share common expressions of *ethnocentrism*; that is, the belief that one's in-group is superior to out-groups (Sumner, 1906). Such beliefs manifest as *in-group favouritism* (favouring members of one's group over those of an out-group), cohesion, and solidarity (Adorno et al., 1950). Groups are also characterised by universal processes related to intergroup differentiation based on evaluative stereotypes which highlight differences between in- and out-group members.

SIT suggests that we can have multiple social identities, due to our membership of multiple social groups. Critical to SIT is the notion that we are motivated to connect with groups in order to develop and maintain a positive SI that enhances our self-esteem, and reduces uncertainty (Hogg, 2018). Hence, the core functions of group membership are to preserve and enhance a positive sense of self, and to optimise distinctiveness from other groups (Hogg, 2018). Group membership achieves this through a process that generates in-group favouritism, arising from comparisons between members within the group and those external to the group (Tajfel & Turner, 1979). As such, SI theories are largely concerned with accounting for processes driving '*them and us*' comparisons (Hogg & Reid, 2006).

SIT outlines that social groups have relative status positions to each other (Hogg & Vaughn, 2018). In the comparative process, in-group members are assessed as possessing *superior* characteristics to out-group members. Group comparisons derive from a criterion in which the out-group performs less favourably (Hennigan & Spanovic, 2012). The group comparisons process maximises perceptions of difference and superiority of the in-group, fulfilling the need for positive distinctiveness from other groups (Brown, 2000). This serves to create a sense of homogeneity that can also minimise differences within the in-group (Barrett et al., 2004). Group membership can become compromised by unfavourable comparisons that threaten positive self-esteem (Hogg & Reid, 2006). Self-esteem and SI are therefore inextricably linked (Haslam et al., 2012), with self-esteem being determined by evaluations of group membership. In contrast to personal identity, which is resistant to change, SI is more malleable, allowing for individuals to transition between identities, and leave groups when discrepancies between the self and group-concept arise.

Such comparative processes are further influenced by (i) whether the person is *subjectively connected* with the group (i.e. their membership has been internalised into their self-concept), (ii) if the *context* of the situation is conducive to comparative evaluations, and (iii) if the comparative group is comparable so as to increase pressure for *distinctiveness*. Studies of street gang behaviour indicate that factors such as how integral a group is to a member's identity, or how distinct the group is, moderates criminal behaviour (Hennigan & Spanovic, 2012).

A key aspect of SIT is the notion that human behaviour is a *continuum* with interpersonal (acting as an individual) and intergroup (acting as a group member) behaviours occupying polar ends. This concept highlights that group behaviour will only manifest under certain conditions, mediated by the permeability, legitimacy, availability, and stability of the group. When operating as an individual, there is a greater variability of behaviours towards in and out-group members (Tajfel, 1978). This is of relevance to those working with forensic populations as it suggests that 'criminal' elements of a social identity are only salient under certain conditions. Furthermore, it suggests that, outwardly, individuals may *behave* in accordance with the group identity, although cognitively and affectively hold differing positions. Therefore, within forensic services, practitioners should not assume the salience of 'criminal' identities, including those with recidivist presentations, at the expense of considering wider factors that contribute to offending behaviour (e.g., coercion from gangs, family members, or peers) which could be of critical importance to contextualising risk assessments. The concept of a 'criminal' identity is explored below.

Self-Categorisation Theory

The key tenants of SIT have been empirically supported in laboratory and naturalistic studies, and it remains one of the most seminal works of social psychology. In a development of the theory, Self-Categorisation Theory (SCT; Turner et al., 1987) was proposed to account for the *cognitive* dimensions of this process. SCT asserts that people define themselves as both individuals and group entities. It encapsulates three core notions, namely that (i) SI is the vehicle that makes group behaviours possible, (ii) self-categorisation is context-dependent, and (iii) when we perceive we have shared similarities with others, we are motivated to cooperate in ways that are consistent with the group's identity (Haslam et al., 2012). SCT also addresses broader questions about how one interfaces with society, including how and when we define ourselves as part of a group (Haslam et al., 2012). As such, SCT offers a more detailed account of the social *cognitive* processes that influence group identification and behaviour.

Key concepts within SCT are social categorisation, depersonalisation, and prototypes. *Self-categorisation* is the process by which people define themselves as being similar or different to classes of stimuli, in *contrast* to each other. Turner et al. (1987) argued that group membership arises as a result of self-categorisations which infer similarities between the self and the 'in-group' and differences with the 'out-group'. Similarly, SIT posits that the consequence of this process is the enhancement of one's self-esteem. While Turner et al. stressed the importance of in-group favouritism as a product of self-categorisation, they also stressed that it is not inevitable (Turner & Reynolds, 2012). When identification with the in-group is salient and membership enhances self-esteem, behaviour will conform to the group's norms (Hopkins & Greenwood, 2013). Key to this process is 'depersonalisation'.

Depersonalisation[2] is defined as "the cognitive redefinition of self" (Turner & Reynolds, 2012), from [individual] unique attributes to [shared/group] social category. Through this process, the group becomes the psychological [shared] reality of its members, and everyday realities are structured by group categorisations (Haslam et al., 2012). Individuals, once categorised, are viewed through the social lens according to their group 'prototype' (Hogg & Reid, 2006).

Cognitive representations of groups are known as *prototypes*. Crudely defined, prototypes are shared social attributions relating to beliefs, feelings, and behaviours that allow group members to identify their own group, and differentiate it from others (Hogg & Vaughn, 2018). Prototypes represent the ideal attitude and behaviour within the group, and include defining and distinguishing features that allow us to infer similarity, coherence, and unity in a process known as entitativity (Hogg & Vaughn, 2018). Hence, our own behaviours and beliefs are prototypical of the in-group with which we identify (Turner et al., 1987).

Prototypes generate stereotypes that provide blueprints to prescribe how we think, feel, and behave towards in and out-group members. Intrinsic to this process is the self-referential nature of comparisons. Hence, self-categorisation has the same depersonalising effect, which allows for the development of self-concept as well as accentuating feelings of contentedness and group identification. Hogg (2018) adds that depersonalisation, at an individual level, also influences how we feel and behave to conform to group (stereotype) norms and prototypes. In essence, through depersonalisation, the self-categorisation processes direct our cognitions, emotions, and behaviours to conform to the group's prototypes.

Through depersonalisation, we filter and interpret social information, including that which is ambiguous, in a prototype-consistent way (Hogg, 2018). Prototypical in-group members are also viewed as the most reliable sources of information when it comes to understanding what constitutes normative or affirming behaviour for the group (Abrahams & Hogg, 1990), reinforcing normative group behaviour.

Both SIT and SCT acknowledge that we can hold memberships to multiple social categories and groups, and that some identities have salience over others. While we hold a number of different social identities, such identities are more likely to be salient when they are habitually available, situationally accessible, and offer both structural and normative fits. This concept of psychological salience is central to understanding how group membership results in depersonalised behaviours in context-dependent ways (Hogg, 2018). The relationships between our different social identities, and the salience of some identities over others, is of key interest to forensic psychology and interventions to reduce criminal identities.

Criminal Social Identity Models

SI theories have continued to evolve to account for conformity (Perfumi, 2020), collective guilt (Ferguson & Branscombe, 2014), deviance (Belmi et al., 2015), group dynamics (Hogg & Reid, 2006), and, most recently, persistent criminal behaviour. The Criminal Social Identity model (CSI model; Boduszek & Hyland, 2011) and the more recent Integrated Psychosocial Model of Criminal Social Identity (IPM-CSI; Boduszek et al., 2016a) offers accounts for the development of a 'criminal social identity' (CSI), formulated as the entrenched expression of a rejected [prosocial] identity.

The models focus on four key processes: (i) psychological crises, (ii) exposure to criminal and dissocial environments, (iii) protection of self-esteem, and (iv) personality traits. These processes contribute pathways to the three factors that constitute a salient CSI, specifically (a) cognitive centrality, which increases conformity to anti-social behaviour through endorsement of group norms, (b) (positive) in-group affect generated by membership, and (c) in-group ties that protect and enhance self-esteem (Spink et al., 2020). Such pathways will now be discussed.

Aetiology of Criminal Social Identity

Psychological Crises

The CSI and IPM-CSI models postulate that criminal identity is, in part, the manifestation of psychological crises in adolescence. Crises are triggered when individuals fail, perceive themselves as rejected, or *are* rejected by peers in their [prosocial] roles, enforcing perceptions of a lower status compared to that of the more successful prosocial group. In turn, this creates a discrepancy between the actual and ideal self, which (i) contributes to goal achievement frustration, (ii) weakens bonds with prosocial groups, and (iii) offers an alternate [criminal] SI that provides a comparatively more ideal sense of self.

Feelings of frustration, which also occur within psychological crises, can be exacerbated by familial factors (Boduszek et al., 2016a). Inappropriate parenting style and rejection, and poor levels of supervision, can prevent the development of both adaptive psychological constructs that protect from offending behaviour, and social bonds that encourage prosocial behaviours. Consequently, these factors can intensify criminogenic associations and cognitions.

Exposure to Criminal Environments and Peers

Incorporating the core principles of Differential Reinforcement Theory (DRT; Akers, 1985), the models also suggest a role for exposure to criminogenic environments and peers (Boduszek et al., 2016a). Exposure to criminal behaviour facilitates the development of the self and other cognitions, values, motivations, and behaviours that increase the likelihood for developing a salient criminal identity. Differential reinforcement allows for gaining knowledge to engage in criminal acts *and* avoid punishment, allowing for the benefits of criminal behaviour to enhance the value of a criminal identity, without the identity being threatened by punishment. This influence increases over time, with prolonged exposure to deviant cognitions and behaviours reinforcing the associations between these variables, especially with in-group ties (Boduszek et al., 2016a).

Protection of Positive Self-Evaluations

Boduszek et al. (2016a) assert that the cognitive centrality of the CSI is to promote positive self-evaluations. The models accept that the exact nature of the relationship between self-esteem and criminal behaviour is complex, and may reflect a mutual reinforcement process whereby those with low self-esteem may initially attract group membership, and that membership can resultantly change self-esteem over time. Continued membership is therefore contingent upon continued positive affirmations. If criminal identity leads to prolonged *negative* self-evaluations, membership could be threatened.

Personality Traits

The models recognise the moderating role of personality traits in the relationship between environmental factors and CSI. Specifically, emphasis is placed on psychopathy, with evidence, albeit limited, for a moderating role of this trait in the formation of a CSI (Boduszek et al., 2013b, 2016b). Additionally, the models suggest a moderating role for introversion and extroversion, with in-group affect having a greater impact on thinking styles of introverted individuals, and in-group ties having a greater impact on criminal thinking in extroverts (Boduszek et al., 2012).

Maintaining Criminal Behaviour

The IPM-CSI model explains persistent displays of criminogenic behaviours as expressions of conformity (to a group identity). Boduszek et al. (2016a) argue that group processes can increase the frequency and range of dissocial behaviours as manifestations of over-conformity. These acts demonstrate the salience of criminal identity, and arise from encouragement; direct pressure is not a prerequisite to maintain the salience of the CSI. Selective affiliation, namely the process of interacting with others in the same category, also reinforces the CSI (Swann, 1987), which may explain why imprisonment- that is, the detainment of individuals with shared criminal identities in a singular facility- is ineffective in reducing recidivism (Harding et al., 2019).

Critique of the CSI and IPM-CSI Models

CSI models have intuitive appeal, and have been subject to empirical evaluation with a narrow range of neurotypical working age, largely (white) male samples, who present without mental disorder. Within this context, evidence, generated largely by Boduszek and colleagues, has supported its main tenants, although it has yet to be applied to service models or intervention frameworks.

At a basic level, research supports the main tenants of the notion of criminal identities and their association with criminal behaviours and arrests. Arrests are positively associated with endorsing a CSI (Boduszek et al., 2014), and identification with antisocial peers also increases the likelihood of developing pro-criminal cognitive structures (Boduszek et al., 2016a), and predicts criminal thinking styles (Boduszek et al., 2013a, 2013b). Incarceration, which exposes 'offenders' to criminal group prototype figures, has been found to increase the likelihood of developing deviant attitudes during custody, particularly in those who present with high levels of interpersonal manipulation (Sherretts et al., 2016). Specifically, the more one engages with criminal peers, the stronger their positive evaluative feelings towards the criminal group become (Boduszek et al., 2016a). This indicates that the length of incarceration, in the presence of psychopathic, manipulative traits, predicts CSI. Adopting a CSI in secure settings could form a survival strategy rather than reflect a 'disordered' identity.

Nonetheless, CSI models were not developed with forensic populations in mind, limiting its utility within such a context. The assumption of CSI being precipitated by a psychological crisis in adolescence has some limitations. For example, while CSI models accept that the development of a CSI might be influenced by representations of known criminals, it fails to consider that this process may lead to individuals bypassing the crisis stage, and developing a CSI through pervasive exposure to criminal behaviour (group prototypes). Indeed, inter- and intra-generational transmission of crime is evident, with high levels of exposure to household violence and incarceration in 'offenders' (Junger et al., 2013). In addition, the impacts of poverty, structural inequalities, and trauma, warrant greater consideration as mechanisms to engaging in criminogenic lifestyles. Therefore, criminal identity may not be the result of a crisis, but of internalising and conforming to an existing salient CSI in their formative developmental environments.

Intergenerational transmission of criminal behaviour and mental health difficulties through disrupted attachment frameworks, is somewhat overlooked by current CSI models. The development of a criminal identity may not be the result of rejection of the group norm, in this case an explicit dissocial norm, but assimilation with it. In essence, a criminal identity may not emerge following a crisis, but instead evolve through assimilating and internalising prevailing models of the self, others, and the (immediate) world. Furthermore, the definition of psychological crises used within the model are relatively narrow, focusing on societal rejection. For example, definitions of rejection in the model fail to consider more nuanced forms of rejection, such as sexual abuse perpetrated by a caregiver.

Current CSI models are also based on the implicit assumption of congruence between public social behaviour, and salient SI. However, some individuals may publicly display group prototype behaviours in order to mobilise group support and avoid their membership being questioned (Klein et al., 2007). As such, identities are fluid, contextually driven, and the perpetual subject of negotiation within the individual; thus, the presented and perceived identity of the individual may not match reality. This phenomenon, known as 'identity performance' (Goffman, 1959), remains in its infancy, despite having potential applicability to forensic populations – especially to trafficked populations, those involved in gang membership, or victims of abuse who are groomed into abusive roles.

Yet, SI theories pose something of a paradoxical dilemma for forensic practitioners. As a meta-theory, SIT has the potential to significantly increase our understanding of the complexity of social identities, rooted in multiple, and often structural, disadvantages that can account for mental disorder and criminal behaviour. That said, the use of labels such as 'criminal social identity' creates ethical challenges for practitioners and academics alike (Becker, 1963). Moreover, formulating offending behaviour within the context of a pervasive CSI generates important ethical debates arising from the effects that a deterministic labelling approach could have. These will now be explored.

Ethical Dilemmas Associated with Social Identity Approaches

The use of SI models in forensics creates a number of ethical dilemmas. Firstly, the language used to describe and categorise an individual can have a profound influence on their place in society, personal identity, and behaviour (Willis et al., 2018). The concept of a pervasive CSI as a framework for understanding complex social behaviours is deeply stigmatising, and further disenfranchises an already marginalised population. While the use of stigmatising labels is commonplace in areas related to mental health, it is particularly salient within forensic services (Willis et al., 2018).

Labels inform the cultural tone, policies, and goals that direct forensic services, and frame the conversations that commissioners, service providers, and health professionals have about those they support. Assigning a 'criminal' SI label exacerbates the idea of 'badness' and 'difference' between those convicted or not convicted of offences, and, in turn, reduces compassion, increases stigma and social alienation, and reinforces in- and out-group comparisons. Labelling individuals as having a CSI also risks the causal use of terms such as 'deviancy' (e.g. Cohn et al., 2015), and the adoption of 'essentialist' views of behaviour (de Vel-Palumbo et al., 2018) as opposed to the recognition of criminal behaviour as social categories that are ameliorable to change.

Relatedly, these approaches increase the likelihood of 'moral' comparisons of individuals, and perceiving criminal behaviour as a (social) rule violation rather than meeting needs (e.g. Cohn et al., 2015). Accumulatively, the attribution of a CSI label could exacerbate the emotive responses of the media and broader society to offending behaviours, with perceptions of individuals reduced to a singular, non-transient identity. It also risks reinforcing the belief that forensic populations cannot be rehabilitated, creating further internal and external barriers to engagement with forensic services.

People who engage in offending often present with pervasive defectiveness, shame, guilt, self-loathing, mistrust, and a lack of self-compassion. A CSI label may exacerbate such beliefs, or, alternatively, become a 'badge of honour' that functions to further reinforce 'difference' from the 'out-group'. Indeed, evidence suggests that labelling an individual as 'criminal' has negative impacts on the development of a prosocial identity, increasing the risk of engaging in future offending behaviour (Willis et al., 2018). Furthermore, the concept of a CSI is incongruent with the current dominant theories in forensic psychology. Within forensic services, reducing perceptions that forensic populations are 'different' has been a key aim of current theoretical models that emphasise moving away from stigmatising labels. As such, reducing complex behaviours to simplistic labels such as a CSI risks regression within the field.

Ultimately, labelling a person's SI as 'criminal' holds a number of important consequences that may, paradoxically, maintain a criminogenic identity and sustain offending behaviour. For SI theories to offer novel, ethical, and compatible frameworks, a refocusing of language is needed. The strength of SI theories may not be in the articulation of a model of 'criminal identity', but in the improvement of understanding the *group* context of offending behaviours, and using this to inform forensic services.

Application of Social Identity Approaches to Forensic Populations

Do SI theories bring anything new to our understanding of offending behaviour and management? For SI theories to be of notable value to forensic psychology, their ability to enhance understanding of complex offending behaviours in the context of mental health needs, while simultaneously ensuring that marginalisation, stigma, and 'difference' are not accentuated, is key. Additionally, they must inform the development of current service delivery and intervention frameworks.

One area of practice that SI approaches could contribute to relates to offending behaviours occurring in the context of group membership. Operating in a group context significantly alters the threshold for offending behaviour, and can lead to more violent offences, and thus more significant injuries for victims (Lantz, 2019). Through collective behaviour processes, including defused responsibility and perceived reduced risk of punishment (Lantz, 2019), escalation of offending behaviour in the context of group-based offending has been observed (Bouchard & Spindler, 2010). Furthermore, group effects appear more prominent in all-male group contexts, and mixed gender groups appear to increase the versatility of female offending (Lantz, 2019). Of particular relevance to forensic psychology is the impact that group processes have on group-based internet offending where anonymity, a diminished sense of responsibility, and escalation of deviancy are particularly prominent. Using the examples of online sexual offending and extremism, we will now outline the potential application of SIT to understanding these behaviours, and delineate how it can be used to inform treatment frameworks and service composition.

Engaging with Online Groups and Extreme Behaviour

The internet can offer sanctuary to those who are rejected, marginalised, or stigmatised in society. Online communities can therefore provide havens of social connection, support, and positive self-esteem. As such, membership to virtual communities can also provide the setting conditions in which group membership leads to engagement in, and escalation of, offending behaviours. In recent years, interest in radicalisation via group membership, as well as sexual offending, has increased. The combined impact of group processes and online offending is likely to be of growing relevance as the role of technology in offending behaviour increases. Online subcultures, particularly for those who have sexual preferences and interests that fall outside of social norms, are a growing concern (Holt et al., 2010). The role of SI theories in increasing our understanding and management of group offending within non-gang affiliated (sexual offending) and 'gang' affiliated (political and religious extremism) groups will now be explored.

Online Sexual Offending Behaviour

Internet-based sexual offending can include the creation and sharing of extreme pornography and pro-offending sites that promote sexual contact between adults and children (Holt et al., 2010). Individuals who are sexually attracted to children represent a highly marginalised group (Holt et al., 2010). These individuals experience high and persistent levels of shame and guilt, which are a chronic threat to self-esteem, and result in both self- and societal-driven segregation. In contrast, membership of online groups allows for accessible sources of pseudo intimacy, belonging, and self-enhancement, which may not be attainable from one's immediate environment (Chiu et al., 2015), and offers access to 'normative' views of sexual behaviour that are consistent with one's own. Thus, online communities can provide a positive [social] identity that becomes a source of self-esteem and social support (Hinduja, 2008).

As well as providing access to normative views, SI theories also account for understanding how barriers can be removed in group scenarios, and how this may escalate into viewing more extreme materials. Group membership can create conditions that remove individual responsibility and internal barriers for *individual* actions. This is especially relevant to internet-based crime, where accessibility and anonymity lowers barriers to extreme or illegal behaviour (Hinduja, 2008). The lowering of internal psychological barriers to offending is achieved through a diminished awareness of self-identity, known as *deindividuation* (Festinger et al., 1952). When deindividuation has occurred, people are more likely to engage in behaviour from which they would normally refrain, as a result of failing to see oneself as an individual (e.g. Paloutzian, 1975).

Traditional conceptualisations of deindividuation (i.e. Le Bon, 1995) postulate that, under group circumstances, an individual loses their personal identity, and inherits a new collective group 'personality' which gives rise to antinormative behaviour. Nevertheless, the principle of 'functional antagonism' – the notion that only one identity can be salient at a given time – has been challenged in more contemporary research, which highlights that multiple identities can be present, but may differ in their saliency (e.g. Rattan et al., 2017). The notion of a dual-identity, in which identities are salient at both an individual- and group-level, may offer important insights for understanding online offending.

The Social Identity Model of Deindividuation Effects (SIDE; Spears & Lea, 1992) suggests that increased awareness of motivations of the group, enhances the salience of the group identity. That is not to say that when placed within a group, a person will 'lose' their personal identity, but rather that the collective group identity becomes more dominant, through deindividuation. When a group identity is more salient, an individual's behaviour is more heavily directed by the goals of the group. Given that anonymity enhances deindividuation, which in turn increases self-awareness as a means for promoting individuation and reducing the salience of the collective group identity, it is probably an important mechanism in desisting from online offending behaviour.

Extremism

Another mechanism that has evolved from this dual-identity perspective, and which may be particularly relevant to understanding displays of *extremist* online behaviour, is 'identity fusion' (Swann & Buhrmester, 2015). SI theories acknowledge a role for *pluralistic ignorance*, in which there is an incongruence between public social behaviour and privately held social norms. Under *normative* social influence, individuals show public conformity to a group for the purpose of acceptance (Myers, 2008). Yet, the particularly strong identification with one's group, which stems from high self-uncertainty, may lead to the private internalisation of one's group membership into one's self-concept, through *informational* social influence (Kelman, 1958). As an individual begins to identify with extremist groups, this new radicalised group identity may overshadow their personal identity, leading to the uncompromising acceptance that presents, even in the presence of more attractive alternatives. Given that personal identity is more resistant to change, 'identity fusion' may explain the high rates of recidivism often seen in extremist offending populations (e.g. Altier et al., 2019). 'De-fusion' may therefore be an important mechanism for desistance from extremist offending behaviours.

Group membership also offers an account for escalation in the extremity of online behaviour. As members become accepted into a group, they are judged more harshly for acts that deviate from group norms. Deviance by a prototypical group member therefore poses a greater threat to both their own identity and that of the group (Pinto et al., 2010). As such, the motivation to engage in anti-social behaviours, particularly those of a greater extremity, is likely to increase as an individual becomes a more established member of the in-group, in order to maximise intragroup similarity and intergroup difference ('*meta-contrast principle*'). Displays of more extreme behaviours by more established and prototypically-central group members, are also likely to shift prototypes towards more socially-polarised norms within the group. Thus, this encourages more extreme behaviours by peripheral members, in an attempt to establish themselves as group members.

A further SI mechanism pertinent to understanding extreme behaviour is uncertainty. Uncertainty-identity theory (Hogg, 2000) postulates that, through SI and self-categorisation processes, feelings of self-uncertainty lead to assimilation with social groups as a way to reduce such feelings. One way through which an individual may seek out uncertainty reduction is through identification with groups, particularly those which are of 'high-entitativity' [seen as being a coherent entity] and have an explicit behavioural prototype. Applying this to the context of extremist behaviours, Hogg (2007) suggests that high uncertainty leads to stronger identification with groups, particularly those of a totalitarian nature. Groups which uphold extremist views and behaviours offer guaranteed entitativity and clear behavioural expectations, which lends to a stronger sense of identity, and thus may be more appealing when self-uncertainty is high. Individuals who are ostracised from society may be at a greater risk for radicalisation, as a result of their heightened need for belonging.

Future Directions: Early Trauma and Social Alienation as the Foci for 'Outgroup' Status

Despite their relevance, SI theories, to date, have had a limited impact on our approaches to understanding and managing offending behaviour (Hennigan & Spanovic, 2012). Current singular SI models offer some value but fail to provide comprehensive accounts to understand the complexities of offending behaviour in individuals with mental health difficulties. In addition to the ethical issues raised above, this approach risks complex forensic needs being formulated in simplistic and reductionist frameworks. Extending current models and integrating existing SI models may offer comprehensive frameworks to forensic practitioners, and a fruitful area for future research. Individuals within the forensic population may present with multiple salient social identities that could indirectly impact on offending behaviour. Nevertheless, research to substantiate this is needed, with notable consideration of social marginalisation resulting from exposure to early and persistent trauma, and substance misuse, being crucial. Forensic populations represent some of the most marginalised and disenfranchised sections of society. They typically face a multitude of structural, social, and economic disadvantages, which interact to perpetuate social exclusion, and maintain a cycle of offending and social alienation (Fergusson et al., 2004). As such, the early adverse experiences of this population may offer an alternative and important framework for understanding the roots of their 'outgroup' identity status and resulting needs, compared to that offered by existing SI models. The relevance of such a lens will now be discussed.

Forensic populations are more likely to have experienced a greater number of Adverse Childhood Experiences (ACEs), including direct abuse, household adversities (e.g., Ford et al., 2019), and entering the 'looked after' care system (Prison Reform Trust, 2016). Those who offend also tend to have received a poorer quality of education, and are more likely to have neurodevelopmental disorders (ibid.). As such, they may not possess the literacy and academic skills needed to thrive in society. Cumulatively, forensic populations' early experiences with family life and prosocial (statutory) agencies, typically include rejection, abuse, inconsistent care, and a failure to have been protected. As such, forensic populations have limited opportunities to develop 'prosocial' identities. While SI models describe early negative experiences as leading to 'psychological crises' that precipitate criminal identities, they perhaps play a more fundamental role in defining a more pervasive outgroup identity. As such, while the language of 'criminal' SI models is questionable, the early adversity, social alienation, and rejection that forensic populations experience described by SI theories, which profoundly shape identity and an 'out-group' status, has significant promise.

Such early adversities create a legacy of enduring attachment and mental health difficulties and economic disadvantage, in the context of social marginalisation and mistrust of prosocial groups. Indeed, people convicted of criminal offences demonstrate greater attachment difficulties (Ogilvie et al., 2014), and increased risk of substance misuse (Abracen et al., 2017), mental health diagnoses, personality disorders, and severe mental illness (Mundt & Baranyi, 2020). As such, by virtue of their diagnoses and offending status, this population risk facing double discrimination, which can also serve to perpetuate isolation. Resultantly, the power of SI models may lie in describing the (social) setting conditions that lead to a *social alienation* (outgroup) status, rather than in labelling individuals. This approach would acknowledge systemic/social contributions, rather than pathologising the individual.

The development of an SI model that focuses on 'what happened to you' (socially marginalised, disenfranchised, and rejected) rather than 'what is wrong with you' (criminal social identity), would be a timely development for forensic populations. Moreover, such a model would stress the role of trauma and marginalisation in pathways to offending, rather than rule violation-based schemas, framing more compassion-based approaches, and reducing stigmatisation. Emphasis on the impact of social marginalisation would also be highly compatible with the principles of trauma-informed care (TIC; Substance Abuse and Mental Health Services Administration, 2014), which are being increasingly recognised as central to meeting the needs of forensic populations. Diverting focus to describing how the marginalised and disenfranchised experiences of this population shape identities and contribute to the formation of alternate [disenfranchised] group membership, lends to the acknowledgment of offending as a response/solution to social status, which would also be consistent with the Good Lives Model (Ward, 2002). Additionally, SI models would add novelty to existing TIC models by articulating how group processes contribute and maintain social alienation and isolation, and outlining what mechanisms services can implement to reduce this.

The enduring mental health needs of forensic populations highlight further benefits of developing a more inclusive framework of 'social alienation' as a pertinent identity, as mental health needs also generate salient identities which impact on engagement with services and remediation attempts. For example, evidence suggests that substance use, which is highly prevalent in forensic populations, can give rise to a valued 'user' SI and the associated cognitive, affective, and behavioural sequalae, which is a powerful determinant of relapse into substance use behaviour (Dingle et al., 2015). Hence, individuals within forensic services may self-identify with multiple salient identities, including 'offender', 'victim', and 'user', as well as additional SI's derived from stable demographic factors that may account for offending behaviour, which, altogether, may be more usefully formulated as a broader social alienation identity.

The benefits of re-directing focus to the common experiences of the salient social [alienation] identities present in forensic populations, does hold some premise, as illustrated in the context of substance use. Substance use has been postulated as a means for the formation of a new identity and sense of belonging in individuals for whom experiences of historical trauma have blocked the development of a pro-social identity (Skewes & Blume, 2019). Studies exploring addiction cessation suggest that the development of (new) social identities, based on cessation, positively impact recovery. Frings and Albery (2015) report that recovery is strengthened when participants have a more positive evaluation of the new 'recovery' identity, relative to their 'addiction' identity. Changes to endorsements of 'recovery' vs. 'user' identities during the course of intervention have been reported, suggesting that SI measures are sensitive to change, increasing their clinical utility. This model has a number of potential applications to forensic work, especially in its emphasis on social support provided by other members of the 'recovery' identity social group. As such, SI models provide a strong theoretical foundation on which to support the expansion and evaluation of initiatives such as Circles of Support and Accountability, which seek to support individuals convicted of sexual crimes to develop prosocial networks, and reintegrate into society. SI models also provide a useful framework to account for the mechanisms of such programmes, particularly in reducing in- and outgroup perceptions, and modelling alternative prototypes.

The development of a theoretically integrative model that accommodates multiple salient social identities as direct and indirect behaviours, and focuses on communicating the impact of exposure to factors that increase the risk of offending, rather than labelling SI on *one aspect* of the resultant behaviour, is likely to have a greater impact on the development of forensic services.

Organisational Implications for Forensic Services

At a fundamental level, SI theories assert that the overarching principles of forensic care are the acknowledgment of the social (group) context of offending behaviour, due to the powerful impact that groups have on cognitive views of the self, others, and world, as well as on behaviour (Hennigan & Spanovic, 2012). This has several key implications for the design and delivery of forensic services. Principally, SI models suggest that focusing treatment through specialist psychological and pharmacological therapies that focus on the individual, are, in themselves, incomplete mechanisms of change, as they fail to address social (group) processes. As such, SI models offer useful insights into service composition and the impact this can have on engagement and outcomes. Some of these potential applications will now be considered.

Team Composition

Forensic mental health teams consist of healthcare professionals who, to service users, are members of the 'out-group'. SI asserts that health professionals could be considered unreliable sources of information. Accordingly, SI models reinforce the importance of initiatives, such as peer workers, to improve therapeutic alliance and treatment engagement. In line with the meta-contrast principle, composing a team of staff who, from the perspective of service users, are more closely aligned with their criminal group norms, challenges the social categorisation of staff as an 'out-group'. In parallel, peer support workers also serve to challenge the central 'criminal' prototype that is dominant within forensic services, offering personal narratives of desistance from offending to promote the benefits of an alternative, prosocial identity. The peer support worker role also offers an alternative identity for those in the position themselves, distinct from their other identities as (ex- or current) 'patient' or 'criminal' (Barrenger et al., 2020).

Staff Conflict, Performance, and Retention

SI approaches may also offer novel insights into dynamic processes that underpin inter-professional teamwork, especially conflicts between workers (Bochatay et al., 2019). Research has demonstrated the impact of team conflicts on the quality of patient care delivered within healthcare settings (e.g. Cullati et al., 2019), and thus it represents an important opportunity for applying a SI approach within forensic organisations.

In line with SCT, perceived similarities and differences between staff working within an organisation lead to the categorisation of others as members of in- and out-groups. Bochatay et al. (2019) make the case for power differentials between professions as the most salient social category within intergroup conflict, leading an individual to question their membership, magnifying negative perceptions of out-groups, and exacerbating discord within an organisation. Such power differentials may play out in scenarios where supervisors fail to fulfil their duties to their subordinates, for example through a lack of support, or inappropriate delegation. In this instance, hierarchical differences do not advantage the in-group (subordinate staff), and thus lead to disillusionment and negative perceptions of the out-group (supervisors).

To minimise intergroup conflict and maximise staff cohesion, a 'team identity' is important (Bochatay et al., 2019). A collectivistic organisational culture is one in which members prioritise the needs of others and behave in accordance with shared group norms, shifting from 'I' to 'we'. A collectivist work identity is an important driver for engagement in 'organisational citizenship behaviours' (OCBs) – discretionary behaviours that come without direct reward, but can promote the effectiveness of an organisation (Organ, 1988). Establishing a collective team identity may also improve staff retention, with members experiencing a sense of 'loss' to one's goals when leaving the workplace.

Enhancements in collaborative working, productivity, and continuity of staff also lend to an improved quality of care for service users. As such, unifying staff through shared goals may hold important benefits at a service-user, staff, and organisational level. One such way to achieve a more collectivist 'team' identity might be through the embedding of a transformational leadership culture. Staff who work under transformational leadership develop strong in-group feelings and a sense of organisational 'allegiance' (Cheng et al., 2016). This collective group identification mediates the association between a transformational leadership culture and OCBs (Tse & Chiu, 2014). Emphasising the shared organisational vision may also reflect a key mechanism in fostering a collectivist 'team' identity. Uniting staff through a common goal reduces perceptions of 'difference' between the in-group and out-groups.

Summary and Conclusion

The self-referential question of 'who am I?' is, in part, answered by our group membership. Over the past forty years, SI theory has developed into the most significant model of group processes (Hogg & Vaughn, 2018). In bridging the gap between 'I' and 'us', SI theories pose something of a dialectical dilemma for forensic practitioners. Such theories can offer frameworks that account for pathways into offending behaviours, as well as their maintenance and desistance. Consequently, they have the potential to significantly increase our understanding of the complexity and multiplicity of social identities, rooted in multiple, and often structural, disadvantages that can account for mental disorder and criminal behaviour. That said, the concept of a pervasive 'criminal' social identity is deeply stigmatising, and could further disenfranchise an already marginalised population. To evolve in line with current theoretical positions, adaptation and expansion of the models is needed. Accordingly, the strength of SIT may not lie in the development and application of a single 'forensic' identity model, but in a model that emphasises understanding offending behaviours, in the context of mental health disorders, as reflecting 'trauma response' or 'disenfranchised' social identities. Accumulatively, an integrated SI approach which stresses a renegotiation of identity by focusing on social inclusion and recovery, rather than reducing rule violation and desistance of offending behaviour, may be key in the design of forensic services, and their success in reducing offending behaviour.

Notes

1 For the purposes of this chapter, the terms social category and social group will be used interchangeably.
2 Depersonalisation in SCT is not to be confused with broader psychiatric definitions that emphasise the deprivation or divesting of individuality; or with sense of detachment and 'unreal' quality to thoughts and experiences.

References

Abracen, J., Looman, J., & Ferguson, M. (2017). Substance abuse among sexual offenders: Review of research and clinical implications. *Journal of Sexual Aggression*, 23(3), 235–250.

Abrahams, D., & Hogg, M. A. (1990). *Social identity theory: Constructive and critical advances*. Springer-Verlag Publishing.

Adorno, T. W., Frenkel-Brunswik, E., Levinson, D. J., & Sanford, R. N. (1950). *The authoritarian personality*. Harpers.

Akers, R. L. (1985). *Deviant behavior: A social learning approach* (3rd edition). Wadsworth.

Altier, M. B., Leonard Boyle, E., & Horgan, J. G. (2019). Returning to the fight: An empirical analysis of terrorist reengagement and recidivism. *Terrorism and Political Violence*, 33(4), 836–886.

Barrenger, S. L., Maurer, K., Moore, K. L., & Hong, I. (2020). Mental health recovery: Peer specialists with mental health and incarceration experiences. *American Journal of Orthopsychiatry*, 90(4), 479–488.

Barrett, M., Lyons, E., & del Valle, A. (2004). The development of national identity and social identity processes: Do social identity theory and self-categorization theory provide useful heuristic frameworks for developmental research? In M. Bennett & F. Sani (eds), *The development of the social self* (pp. 159–188). Psychology Press.

Becker, H. S. (1963). *Outsiders: Studies in the sociology of deviance*. Free Press.

Belmi, P., Barragan, R. C., Neale, M. A., & Cohen, G. L. (2015). Threats to social identity can trigger social deviance. *Personality and Social Psychology Bulletin*, 41(4), 467–484.

Bochatay, N., Bajwa, N., Blondon, K., Junod Perron, N., Cullati, S., & Nendaz, M. (2019). Exploring group boundaries and conflicts: A social identity theory perspective. *Medical Education*, 53(8), 799–807.

Boduszek, D., Adamson, G., Shevlin, M., & Hyland, P. (2012). The role of personality in the relationship between criminal social identity and criminal thinking style within a sample of prisoners with learning difficulties. *Journal of Learning Disabilities and Offending Behaviour*, 3(1), 12–23.

Boduszek, D., Adamson, G., Shevlin, M., Hyland, P., & Bourke, A. (2013a). The role of criminal social identity in the relationship between criminal friends and criminal thinking style within a sample of recidivistic prisoners. *Journal of Human Behavior in the Social Environment*, 23(1), 14–28.

Boduszek, D., Shevlin, M., Adamson, G., & Hyland, P. (2013b). Eysenck's personality model and criminal thinking style within a violent and nonviolent offender sample: Application of propensity score analysis. *Deviant Behavior*, 34(6), 483–493.

Boduszek, D., Dhingra, K., & Debowska, A. (2016a). The Integrated Psychosocial Model of Criminal Social Identity (IPM-CSI). *Deviant Behavior*, 37(9), 1023–1031.

Boduszek, D., Dhingra, K., & Debowska, A. (2016b). The moderating role of psychopathic traits in the relationship between period of confinement and criminal social identity in a sample of juvenile prisoners. *Journal of Criminal Justice*, 44, 30–35.

Boduszek, D., & Hyland, P. (2011). The theoretical model of criminal social identity: Psychosocial perspective. *International Journal of Criminology and Sociological Theory*, 4(1), 604–614.

Boduszek, D., O'Shea, C., Dhingra, K., & Hyland, P. (2014). Latent class analysis of criminal social identity in a prison sample. *Polish Psychological Bulletin*, 45(2), 192–199.

Bouchard, M., & Spindler, A. (2010). Groups, gangs, and delinquency: Does organization matter? *Journal of Criminal Justice*, 38(5), 921–933.

Brown, R. (2000). Social identity theory: Past achievements, current problems and future challenges. *European Journal of Social Psychology*, 30(6), 745–778.

Cheng, C., Bartram, T., Karimi, L., & Leggat, S. (2016). Transformational leadership and social identity as predictors of team climate, perceived quality of care, burnout and turnover intention among nurses. *Personnel Review*, 45(6), 1200–1216.

Chiu, C., Huang, H., Cheng, H., & Sun, P. (2015). Understanding online community citizenship behaviors through social support and social identity. *International Journal of Information Management*, 35(4), 504–519.

Cohn, A., Maréchal, M. A., & Noll, T. (2015) Bad boys: How criminal identity salience affects rule violation. *The Review of Economic Studies*, 82(4), 1289–1308.

Cullati, S., Bochatay, N., Maître, F., Laroche, T., Muller-Juge, V., Blondon, K., Perron, N. J., Bajwa, N., Vu, N., Kim, S., Savoldelli, G., Hudelson, P., Chopard, P., & Nendaz, M. (2019). When team conflicts threaten quality of care: A study of health care professionals' experiences and perceptions. *Mayo Clinic Proceedings*, 3(1), 43–51.

de Vel-Palumbo, M., Howarth, L., Brewer, M. B. (2018). 'Once a sex offender always a sex offender'? Essentialism and attitudes towards criminal justice policy. *Psychology, Crime and Law*, 25(2), 1–40.

Dingle, G. A., Cruwys, T., & Frings, D. (2015). Social identities as pathways into and out of addiction. *Frontiers in Psychology*, 6, 1795.

Ferguson, M. A., & Branscombe, N. R. (2014). The social psychology of collective guilt. In C. von Scheve & M. Salmela (eds), *Collective emotions: Perspectives form psychology, philosophy, and sociology* (pp. 251–265). Oxford University Press.

Fergusson, D., Swain-Campbell, N., & Horwood, J. (2004). How does childhood economic disadvantage lead to crime? *The Journal of Child Psychology and Psychiatry*, 45(5), 956–966.

Festinger, L., Pepitone, A., & Newcomb, T. (1952). Some consequences of deindividuation in a group. *Journal of Abnormal and Social Psychology*, 47, 382–389.

Ford, K., Barton, E. R., Newbury, A., Hughes, K., Bezeczky, Z., Roderick, J., & Bellis, M. A. (2019). *Understanding the prevalence of adverse childhood experiences (ACEs) in a male offender population in Wales: The prisoner ACE survey*. Bangor University.

Frings, D., & Albery, I. P. (2015). The social identity model of cessation maintenance: Formulation and initial evidence. *Addictive Behaviors*, 44, 35–42.

Goffman, E. (1959). *The presentation of the self in everyday life*. Doubleday Anchor.

Harding, D. J., Morenoff, J. D., Nguyen, A. P., Bushway, S. D., & Binswanger, I. A. (2019). A natural experiment study of the effects of imprisonment on violence in the community. *Nature Human Behaviour*, 3, 671–677.

Haslam, S., Reicher, S., & Reynolds, K. (2012). Identity, influence, and change: Rediscovering John Turner's vision for social psychology. *British Journal of Social Psychology*, 51(2), 201–218.

Hennigan, K., & Spanovic, M. (2012). Gang dynamics through the lens of social identity theory. In F. A. Esbensen, & C. L. Maxson (eds), *Youth gangs in international perspective* (pp. 127–149). Springer.

Hinduja, S. (2008). Deindividuation and internet software piracy. *Cyberpsychology & Behavior*, 11(4), 391–398.

Hogg, M. A. (2000). Subjective uncertainty reduction through self-categorisation: A motivational theory of social identity processes. *European Review of Social Psychology*, 11, 223–255.

Hogg, M. (2007). Uncertainty-identity theory. In M. P. Zanna (ed.), *Advances in Experimental Social Psychology* (pp. 69–126). Elsevier.

Hogg, M. A. (2018). Social identity theory. In P. J. Burke (ed.), *Contemporary social psychological theories* (pp. 112–138). Stanford University Press.

Hogg, M. A., & Reid, S. A. (2006). Social identity, self-categorization, and the communication of group norms. *Communication Theory*, 16(1), 7–30.

Hogg, M., & Rinella, M. (2019). Social identities and shared realities. *Current Opinion in Psychology*, 23, 6–10.

Hogg. M., & Vaughn, G. (2018). *Social psychology* (8th edition). Pearson.

Holt, T., Blevins, K., & Burkert, N. (2010). Considering the pedophile subculture online. *Sexual Abuse: A Journal of Research and Treatment*, 22(1), 3–24.

Hopkins, N., & Greenwood, R. (2013). Hijab, visibility and the performance of identity. *European Journal of Social Psychology*, 43(5), 438–447.

Junger, M., Greene, J., Schipper, R., Hesper, F., & Estourgie, V. (2013). Parental criminality, family violence and intergenerational transmission of crime within a birth cohort. *European Journal on Criminal Policy and Research*, 19(2), 117–133.

Kelman, H. C. (1958). Compliance, identification, and internalization: Three processes of attitude change. *Journal of Conflict Resolution*, 2(1), 51–60.

Klein, O., Spears, R., & Reicher, S. (2007). Social identity performance: Extending the strategic side of SIDE. *Personality and Social Psychology Review*, 11(1), 28–45.

Lantz, B. (2019). Co-offending group composition and violence: The impact of sex, age, and group size on co-offending violence. *Crime & Delinquency*, 66(1), 93–122.

Le Bon, G. (1995). *The crowd: A study of the popular mind*. Transaction.

Mundt, A. P., & Baranyi, G. (2020). The unhappy mental health triad: Comorbid severe mental illnesses, personality disorders, and substance use disorders in prison populations. *Frontiers in Psychiatry*, 11, 804.

Myers, D.G. (2008). *Social psychology* (9th edition). McGraw-Hill Companies.

Ogilvie, C. A., Newman, E., Todd, L., & Peck, D. (2014). Attachment and violent offending: A meta-analysis. *Aggression and Violent Behavior*, 19(4), 322–339.

Organ, D. W. (1988). *Organizational citizenship behavior: The good soldier syndrome*. Lexington Books.

Over, H. (2016). The origins of belonging: Social motivation in infants and young children. *Philosophical Transactions of the Royal Society B: Biological Sciences*, 371, 1–8.

Paloutzian, R. F. (1975). Effects of deindividuation, removal of responsibility, and coaction on impulsive and cyclical aggression. *The Journal of Psychology*, 90, 163–169.

Perfumi, S. C. (2020). Social identity model of deindividuation effects and media use. In J. Van den Bulck (ed.), *The international encyclopedia of media psychology*, (pp. 1–8). John Wiley & Sons.

Pinto, I. R., Marques, J. M., Levine, J. M., & Abrams, D. (2010). Membership status and subjective group dynamics: Who triggers the black sheep effect? *Journal of Personality and Social Psychology*, 99(1), 107–119.

Prison Reform Trust. (2016). *In care, out of trouble*. Prison Reform Trust.

Rattan, A., Steele, J., & Ambady, N. (2017). Identical applicant but different outcomes: The impact of gender versus race salience in hiring. *Group Processes & Intergroup Relations*, 22(1), 80–97.

Sherif, M. (1966). *Group conflict and co-operation: Their social psychology*. Psychology Press.

Sherretts, N., Boduszek, D., & Debowska, A. (2016). Exposure to criminal environment and criminal social identity in a sample of adult prisoners: The moderating role of psychopathic traits. *Law and Human Behaviour*, 40(4), 430–439.

Skewes, M. C., & Blume, A. W. (2019). Understanding the link between racial trauma and substance use among American Indians. *American Psychologist*, 74(1), 88–100.

Spears, R., & Lea, M. (1992). Social influence and the influence of the 'social' in computer-mediated communication. In M. Lea (ed.), *Contexts of computer-mediated communication* (pp. 30–65). Harvester Wheatsheaf.

Spink, A., Boduszek, D., & Debowska, A. (2020). Investigating the Integrated Psychosocial Model of Criminal Social Identity (IPM-CSI) within a sample of community based youth offenders. *Journal of Human Behavior in the Social Environment*, 30(2), 118–137.

Stets, J. E., & Burke, P. J. (2000). Identity theory and social identity theory. *Social Psychology Quarterly*, 63(3), 224–237.

Substance Abuse and Mental Health Services Administration. (2014). SAMHSA's concept of trauma and guidance for a trauma-informed approach. Retrieved from https://store.samhsa.gov/sites/default/files/d7/priv/sma14-4884.pdf.

Sumner, W. G. (1906). *Folkways: A study of the sociological importance of usages, manners, customs, mores and morals*. Ginn and Co.

Swann, W. B. (1987). Identity negotiation: Where two roads meet. *Journal of Personality and Social Psychology*, 53(6), 1038–1051.

Swann, W. B., & Buhrmester, M. (2015). Identity fusion. *Current Directions in Psychological Science*, 24(1), 52–57.

Tajfel, H. (1978). *Differentiation between social groups: Studies in the social psychology of intergroup relations*. Academic Press.

Tajfel, H., & Turner, J. C. (1979). An integrative theory of intergroup conflict. In W. G. Austin, & S. Worchel (eds), *The social psychology of intergroup relations* (pp. 33–37). Brooks/Cole.

Tajfel, H., & Turner, J. C. (1986). The social identity theory of intergroup behavior. In S. Worchel, & W. G. Austin (eds), *Psychology of intergroup relation* (pp. 7–24). Hall Publishers.

Tse, H. H., & Chiu, W. C. (2014). Transformational leadership and job performance: A social identity perspective. *Journal of Business Research*, 67(1), 2827–2835.

Turner, J. C., Hogg, M. A., Oakes, P. J., Reicher, S. D., & Wetherell, M. S. (1987). *Rediscovering the social group: A self categorization theory*. Blackwell.

Turner, J. C., & Reynolds, K. J. (2012). Self categorization theory. In P. A. Van Lange, A. W. Kruglanski, & E. T. Higgins (eds), *Handbook of theories of social psychology: Volume 2* (pp. 399–417). Sage Publications.

Ward, T. (2002). Marshall and Barabaree's intergrated theory of child sexual abuse: A critique. *Psychology, Crime & Law*, 8(3), 209–228.

Willis, M. D., Leaitch, W. R., & Abbatt, J. P. D. (2018). Processes controlling the composition and abundance of arctic aerosol. *Reviews of Geophysics*, 56(4), 621–671.

4 Impression Management

Joel Harvey and Deborah H. Drake

> No-one would seriously challenge the general idea that observers infer dispositions from an actor's behavior or that actors have a stake in controlling the inferences drawn about them from their actions.
>
> (Jones & Pittman, 1982, p. 231)

Introduction

Ryan, aged 17, arrives at the reception area in a Young Offender Institution (YOI). He scans the environment as he walks through the gates into the reception area and he notices some other young people in a holding room and several prison officers standing at a desk. He is asked to hand over his belongings that he has brought with him from court and is then taken into another room to be strip searched. Ryan is then taken into the holding cell to wait with some other young people before being escorted onto the accommodation wing where his cell is. Walking through the corridors, with two other young people and prison officers, he stares at the floor. He doesn't want to make eye contact. He is taken to his cell where he meets his cellmate, Billy. At first, Ryan doesn't say much but they soon start talking. Billy, who happens to come from the same area in London as Ryan, starts to offer him some advice. Billy tells him: 'keep yourself to yourself in here'; 'keep your head down and don't cause any trouble'; 'try not to wind up the officers – be polite to them and laugh at their jokes when they make them'. Billy warns 'not to show weakness' and tells Ryan that he needs to stand his ground, or else others will take advantage of him. When Billy asks Ryan what he is in for, Ryan says 'assault' – he had been told by an officer on reception never to reveal that he was convicted of a sexual offence. At his induction, Ryan was advised by the induction officer that if he let others in the prison know his offence, he could end up being beaten or bullied. He was told to make sure he knew what to say if another young person asked.

What stands out here is that, even in the first 24 hours of custody and so long as he remains there, Ryan will need to learn new ways of carefully managing how he is perceived by others. This will require multiple acts of impression management, according to different situations and in

DOI: 10.4324/9781315560243-4

conversations with people who hold different forms of power in relation to Ryan. It will be an arduous task that will require Ryan to be consciously aware of how he is coming across – both to other young people and to staff – in order to navigate his way around the system and, ultimately, survive.

In this chapter, we explore the concept of impression management and how it is enacted in secure forensic settings. We argue that one of the ways impression management can be understood is as a coping mechanism that many human beings innately develop as part of learning how to successfully navigate the social world and remain connected to others. As an adaptive social survival skill, it might be argued that impression management develops as we form, or are threatened with the loss of, social attachments. This does not mean that impression management strategies are always healthy, benign, or adaptive responses (Emery et al., 2014; Flynn et al., 2018). However, if we begin from the premise that impression management first develops as a means of navigating important relationships it allows for a more nuanced understanding of it as not just a conscious decision, but also as an unconscious reflex that emerges in response to perceived threats to our psychological or physical safety, our social status, or our sense of connectedness to others.

To develop our arguments we briefly consider the links between attachment theory and impression management (Emery et al., 2014; Flynn et al., 2018), before going on to consider the complexities of the performance of self (Goffman, 1959) in the high-stakes environment of total institutions (Goffman, 1961). We then move on to consider the concerns that forensic practitioners may have with regards to deception and impression management. We consider the work of Paulhus (1986, 1998 *inter alia*) on (self-)deception to argue that living and working in forensic settings place extraordinary demands on both clients and practitioners who are forced to navigate an artificial, yet often threatening and power-dense, world. Our survey of the research literature describes a complex picture, but one which does not ultimately confirm a straightforward link between (self-)deception, impression management, and criminal attitudes. The chapter then moves on to examine the ideas of the 'public' self and the 'private' self (Tice, 1992) and the different motivations people have for presenting a particular self-image to others or to themselves. This then leads into our discussion of the way self-discrepancy theory illuminates the disjuncture between the 'actual' and 'ideal' self (Higgins, 1987) and the role that shame and guilt can play in influencing self-presentation. Throughout the chapter, we suggest that the social and relational contexts of prisons and other secure settings must always be taken into account when thinking about risk assessment, client engagement, and therapeutic relationality.

Attachment and Impression Management

The impressions others have of us impact upon how they respond to us and how we feel about ourselves (Leary, 2019). We engage (either consciously or unconsciously) in a process of 'self-presentation' or 'impression management' in order to control how we are perceived by others. Think about the last time you were in an online professional meeting via video call. Did you think about what 'background' you would choose? Did you monitor your facial reaction onscreen as someone was making a point with which you did not agree? Did you consider how you wanted to 'position' yourself? In everyday interactions, online or face-to-face, we devote cognitive and emotional energy to attempts to manage how we are seen by others. We can be motivated to portray to others that we *do* possess certain characteristics (and engage in 'attributive tactics') and equally motivated to portray to others that we *do not* possess certain characteristics (and engage in 'repudiate' tactics; Leary, 2019; Roth et al., 1988). But why does what others think of us matter so much? If we consider this question in light of the last several decades of attachment research, one possible answer might be: because on a deep, instinctive level, we feel that our survival depends on our connectedness to others.

The theory of attachment was first developed by John Bowlby (1969) who described attachment as a sense of "lasting psychological connectedness between human beings" (Bowlby, 1969, p. 194). Focusing on infants and their parents, Bowlby considered infant behaviours in relation to parental presence or separation. He found that young children will engage in some quite extreme behaviours (screaming, clinging, crying) to avoid being separated from their parents. Such behaviours are innate or instinctive responses to the perceived threat of losing contact with those who are able to help the child survive (i.e. the caregiver). Such behaviours are part of what Bowlby termed an 'attachment behavioural system' that then guides us throughout our lives to form and maintain relationships. When children form insecure attachments (i.e. they have been unable to consistently rely on caregivers to meet their basic physical or emotional needs), they may develop attachment patterns that also influence and/or impede their adult relationships.

The link between attachment theory and impression management are under-researched in the psychological literature. However, some previous research seems to suggest a link. A few studies, for example, have made links between impression management and attachment theory in the context of social media profiles and behaviours (Emery et al., 2014; Flynn et al., 2018). Emery et al. (2014) argue that: "People attempt to manage others' impressions of them to decrease a disparity between their current images and their desired images" (p. 1467). Part of the reason for doing this is to maintain a desired type of relationship with another. Managing how others view us can help us to facilitate the kinds of connections we have, or want to have, with others. This can include attempting to draw another closer to us or maintain our distance, thus allowing us to affirm or adjust our personal boundaries or try and develop a closer relationship to someone that we like. At the same time, impression management can play a vital role in assisting people to form social networks that can, in turn, help us to build our social capital and connectedness to others in ways that help us succeed (Putnam, 1995). We will develop the ways that attachment theory and impression management might operate in tandem in later sections of this chapter. We now turn to introducing the issues associated with impression management or self-presentation in forensic secure settings.

Defining Self-Presentation in Forensic Settings

Erving Goffman (1959), in his classic anthropological work *The Presentation of Self in Everyday Life*, introduced the idea that interactions with others can be seen as a performance that we undertake, like 'actors' on a 'stage'. Goffman's theory builds on Mead's (1934) symbolic interactionism which argues that the 'self' that people portray to one another is shaped by social relationships and conditions. In essence, Mead saw the self as a 'social product'. Stemming from this, Goffman's theory uses the analogy of the theatre to explore the presentation of self. He argues that social interactions can be analysed by seeing individuals as performers, audience members, and outsiders, acting out scenes on particular 'stages' or settings. He suggests that there is often a division between what might be called the backstage, where the performance is rehearsed and prepared, and a front stage where it is presented. Those giving the performance will control access to these areas to stop the 'audience' from seeing what is going on behind-the-scenes, and to prevent others from seeing a performance not intended for them. As performers, Goffman argues, individuals are concerned with maintaining the impression they have sought to project in answer to the perceived standards against which they are judged.

For Goffman (1959) 'impression management' can be defined as a decision to conceal some aspects of the self, while revealing others. The self emerges as a 'collaborative product' shaped by other actors, the audience, and their interpretive activities. Goffman defined performance as "all the activity of a given participant on a given occasion which serves to influence in any way any of the other participants" (p. 26). He also stated that the image that we portray to others provides them with an idea of how they expect to be treated, and also how they should expect to treat others. Similarly, Leary (2019) defines impression management (which we are using synonymously with self-presentation) as "the process by which individuals attempt to control how they are perceived by other people" (p. 7245). Such 'managed' presentations can influence the response of others. Indeed, it is through the management of impressions within social interactions that expectations are negotiated, and social norms are established.

Goffman (1961) subsequently went on to consider social life in 'total institutions' which are places that are set up to control all or most aspects of the lives of people who are housed within them. Examples of total institutions include: orphanages, boarding schools, hospitals, and prisons. Goffman argued that the experience of being inducted into a total institution was not one of becoming 'acculturated' to the environment. On the contrary, it is a process of disculturation or role-stripping where the person who may have known themselves as a multi-faceted individual on the outside, must now adhere to the binary organisation of the prison, taking their place as an 'inmate' differentiated only in relation to 'officers' or other non-inmate professionals who work within the environment (and who, importantly, can leave every night). This creates a power-dense and socially artificial environment that instantly calls upon the person's most basic coping mechanisms in order to survive. Think back to the short vignette we opened this chapter with and Ryan's experience of learning how to navigate and hone his self-presentation in response to the new world of the YOI. In secure settings, the importance of impression management operates across all relationships. In prisons, for example, prisoners must try to manage how they are seen by other prisoners, which might require a different form of impression management from the way they are seen by prison officers. Likewise, prison officers, psychologists, and other professionals within the prison will all be engaging in their own forms of impression management at various times too.

Forensic clients and forensic practitioners are involved in a self-presentation dance, engaged in managing how they come across to one another. The 'audiences' within forensic settings are varied for both clients and staff. For clients, it can be different professionals with whom they come into contact, including prison officers, nursing staff, solicitors, parole board members, psychologists, psychiatrists (to name but a few), significant others who might write, come to visit, or whom they might telephone; and fellow prisoners or patients. People deliberately engage in impression management when they are concerned with how they come across to others and when it is related to obtaining a valued goal (be that an internal or external goal) or when there is a gap or discrepancy between how the individual believes they are perceived (e.g. vulnerable, threatening) and how they want to be perceived (e.g. resilient, cooperative). Such discrepancies are present across different contexts for clients (and for practitioners too) and can involve managing a myriad of often conflicting presentations to professionals, peers, and significant others. It might be suggested that the context of forensic settings requires people to engage in a kind of *differential adaptive self-presentation*. In practice, this means becoming extremely skilful at enacting an image of self that is coherent, but which can be adapted to the interactional or relational demands of the environment in ways that support the person's social, physical, or psychological survival. In this way, self-presentation can be tactical and based on what a person wants to achieve at a given moment in time (Leary, 2019).

Whether staying in a secure hospital or prison, or undertaking a programme as part of a community sentence, the impressions that are formed about the client can have profound impacts on decisions made about their future, and, on a day-to-day basis, their survival. Forensic clients are under the spotlight and their cognitive, social, emotional, and behavioural 'performance' is being evaluated, often by a collective multi-disciplinary team who have power over the narratives that are constructed about them (in both verbal and written reports) and, ultimately, about their future progression. Such decisions may include whether a client is safe to be granted unescorted leave in the hospital grounds (or whether they might abscond or re-offend), whether a client is safe to have access to a particular belonging (or whether they might harm themselves with it), or indeed, whether someone should be discharged from hospital or released from prison. The client's ability to demonstrate that they *do* possess certain characteristics (such as an ability to regulate their emotions) and that they *do not* possess other characteristics (such as anti-social attributes) could have a profound impact on their future.

Impression Management and Deception

Leary and Kowalski (1990) have argued that impression management is a common-place concern and that "people have an ongoing interest in how others perceive and evaluate them" (p. 34). Leary (2019) also points out that psychologists sometimes view self-presentation "as inherently manipulative and deceptive, the ugly underbelly of interpersonal life" (p. 9). In order to navigate the social world, human beings need to be able to draw on different aspects of the self. This is true for all social settings and for people in all walks of life. But, to what extent can impression management be deceptive? This may be a question that is of particular interest to those working in forensic settings.

When considering deception and what people choose to disclose (or conceal), we would argue that it is important to think about this within the context of the particular relationship through which a version of self is presented. It is important for the practitioner to be aware of the factors that might lead to a particular performance of self-presentation (for example, to achieve a goal or to avoid feeling shame) and that the practitioner needs to deftly attempt to peek behind the mask of self-presentation in order to determine the client's needs. For example, we might ask a young person living in a secure environment: 'How are you?' and the young person replies: 'Fine'. This might be an attempt to present themselves as able to cope and to avoid further questioning. We might follow up with: 'How are you *really*?' to help the young person acknowledge that we do not think they are being deceptive or lying to us but that we know they might not want to expose their underlying feelings. This attempt to try and dig a little deeper, if sensitively done, signals that it is acceptable and permissible for the young person to open up a little more. It is important for the practitioner to attune to the self behind the mask rather than over-relating to the presentation of the mask. There is a complex dynamic at play here – at once not accusing someone of lying but not settling for an initial, terse response. This is also of importance within the therapeutic relationship when focusing on behaviours that may be associated with a high degree of shame for the client. For example, consider the relational dynamics when working with someone who has committed an offence that they are ashamed of (Gilligan, 2003), such as a sexual offence. Rather than acknowledging their offence and opening up about it, it might be psychologically safer for them to deny what happened in an attempt to 'save face' and manage the self publicly. 'Saving face' can be due to both internal processes, such as the shame a person is feeling inside of themselves, as well as their external image, particularly in a prison context where those convicted of sexual offences are often subjected to bullying, violence, and harassment by the 'mainstream' prison population (Drake, 2006). Practitioners or therapists conducting group work (such as the Sex Offender Treatment Programmes) might find it useful to bear in mind that those with whom they are working may be guarding against disclosure in order to avoid shame and/or censure. While there will be incongruence between the private and public self, the perceived benefits of this psychological discrepancy might outweigh the psychological cost of publicly owning the behaviour.

Developing a more nuanced awareness of the functional and, arguably, adaptive reasons that someone may engage in strategic impression management can be a helpful guide for the forensic practitioner to assess the needs of their clients more skilfully and accurately. However, it also remains important for the forensic practitioner to be aware of the strategies of impression management that a client might use during assessments for future violence. While practitioners rely on corroborating information to carry out risk assessments, especially when developing an understanding of historical risk factors for structured professional judgement (SPJ) tools such as the HCR-20 (Douglas & Reeves, 2010), the forensic practitioner also relies heavily on the self-report and disclosure of the client. Much has been written on the area of deception and malingering (Rogers & Bender, 2018) and how some clients might be motivated to present themselves in a manner that will improve their future outcomes. In the context of forensic practice, decisions that are made about a client may have a profound impact on their current circumstances and their future. On the one hand, for the forensic practitioner to do their job and make a clinical judgement, they need to base their decision-making, in part, on self-report, and therefore rely on the client to be honest with them during the assessment. On the other hand, the client is aware of the currency of this assessment and therefore might be guarded in their responses. Such guardedness, however, can result in a presentation of self who seems reluctant to engage, who might lack insight into their difficulties, or who is simply coming across as disingenuous or inauthentic.

Gillard and Rogers (2015) have argued that "deception is a common occurrence in many different settings, especially when the incentives are high […] Nowhere is this observation truer than in forensic settings" (p. 107). Indeed, Gozna et al. (2001) argue that 'deception' is a 'high-stake' behaviour for clients who are undergoing assessments. Previous research has found that forensic clients may not disclose information associated with certain risk factors (Edens & Ruiz, 2006), may conceal 'psychopathic' traits (Edens et al., 2001), hide difficulties managing anger (McEwan et al., 2009), and difficulties with substance misuse (Knight et al., 2002). The issue of deception also needs to be contextualised against the backdrop of the environment in which it is taking place. For example, can a detained person ever truly consent to treatment, particularly in an environment where engagement can determine freedom and where confidentiality does not exist (Simms-Sawyers et al., 2020)? These unspoken 'ethics' and the hidden power dimensions that flow and operate in closed environments should also be considered when seeking to understand why someone might avoid disclosure and/or engage in acts of deception (on the complexities of disclosure in forensic settings, see Ort, 2020).

Williams, Rogers, and Hartigan (2020) argue that forensic clients "may be strongly motivated to appear non-dangerous when administering risk assessment measures" (p. 83) and discuss how 'risk minimisation' is a form of positive impression management. In their study, they asked participants who were in a court mandated residential treatment programme for substance misuse problems, to minimise or deny risk, and found that, overall, the Psychology Inventory of Criminal Thinking Styles (PICTS; Walters, 1995) was robust against impression management, but that participants sometimes under-estimate impulsive and entitled attitudes on this measure.

Gillard and Rogers (2015) examined the effects of positive impression management on risk assessments carried out with people labelled as 'psychopathic' compared to those who did not meet criteria for this label. They argued that "psychologists have a vested interest in developing strategies for risk assessment that are not susceptible to positive impression management" (p. 106). Indeed, Book et al. (2006) found that those who score higher on a measure of deception (using the Holden Psychological Screening Inventory) score higher on psychopathy. Gillard and Rogers sought to explore this phenomenon further and, in their research, established two conditions: an 'honest condition' where respondents were assured confidentiality, and an 'experimental condition' where the respondents were told to pose as though they were at low risk of re-offending. They found that factor 1 of the PCL-R (Hare, 1998), which measures interpersonal and affective traits, was most predictive of deception. It was concluded that risk assessments (including the HCR-20 which was used in this study) are susceptible to positive impression management, thus suggesting that practitioners may need to be mindful of positive impression management techniques when undertaking risk assessment work, but that the research literature remains inconclusive.

Tully and Bailey (2017) examined the role of impression management in relation to personality traits. They used the BDIR-7 (Paulhus 1998; which is also known as the Paulhus Deception Scale; PDS). In particular, they used the impression management sub-scale (recognising shortcomings but then 'faking good') and the self-deception enhancement sub-scale to ascertain if there were any associations with anti-social personality traits and narcissistic traits in a non-clinical UK sample. It was found that those who scored higher on the impression management subscale were *negatively* associated with antisocial traits, and that the self-deception sub-scale was *positively* associated with narcissistic traits. This latter finding may provide some evidence for theorisations that postulate an underpinning role of shame in the manifestation of narcissistic traits (e.g. Nathanson, 1987). Tully and Bailey (2017) conclude that when carrying out assessments with clients with antisocial traits, it should not be assumed that clients who score highly on impression management scales are being antisocial, and it should not invalidate other assessment scores.

Uziel (2010) has examined whether impression management scales are reliable in detecting socially desirable responses, "a tendency by respondents to portray on overly positive image of their true selves" (p. 243). From his review, he concludes that the scales are "less than perfect" (p. 247) and argues that impression management scales might be better at measuring aspects of personality. Based on the literature, Uziel identifies two underlying functions of scoring positively on impression management: *defensiveness approach* and *adjustment approach*. The defensiveness approach to impression management is an attempt to guard against social rejection and threat. Those scoring high on impression management scales are not attempting to achieve social *approval*, but are attempting to actively avoid social *disapproval*. The adjustment approach to impression management is an attempt to create harmony, belongingness, and enhance wellbeing. Uziel argues that this approach involves 'interpersonally oriented self-control' to achieve harmonious relationships, and demonstrates agreeableness and conscientiousness. These individuals are thought to be exercising self-regulation because they want to do 'the right thing' but not in an attempt to manipulate. Such a stance is contrary to the view that a high impression management score indicates an individual attempting to consciously deceive and is in keeping with arguments we have put forward earlier in the chapter about the role impression management plays as a part of skilful and common-place social interactions. Furthermore, understanding impression management as an everyday social psychological activity may also explain why high scores on impression management are not always associated with criminal attitudes (Mills & Kroner, 2006).

The extent to which people deliberately focus on managing their self-presentation can also depend on whether someone is in a public or private realm (Leary, 2019). When considering both clients and practitioners involved in different aspects of the criminal justice system, it is worth remembering that the 'private realm' is almost always a 'public' realm and, as a result, the motivational pressure to self-monitor increases. While a person in prison will spend substantial time within the 'private' (albeit in a public building) space of their prison cell, the minute they set foot outside of their room they are in a public space. Time is then spent interacting with fellow prisoners and with staff, where the need to manage impressions is ever present. For example, when a young person is out on 'association', impression management strategies might be highly engaged in, in order to survive. The young person will, consciously or unconsciously, be thinking about how they are coming across to fellow prisoners and whether this is at odds or in congruence with the 'inmate code' (Harvey, 2012; Ugelvik,

2016). They will need to think about how they are coming across to staff in an offending behaviour programme or when interacting with a prison officer to ask whether it's possible to make a phone call. How does the young person manage their self-presentation when their loved ones come to visit? What strategies are then enacted? Does the young person present as someone who is coping with the prison experience to allay concerns of a mother who might be distressed, or do they present as vulnerable and in need of emotional support? Within the complex environment of the prison, the young person will be managing multiple impressions that may contradict one another. Indeed, how a young person manages his self-presentation towards an officer might be different to how he manages the impression he attempts to give to a fellow prisoner. Thus, the forensic space blurs that of the private and the public. It is a space in which private domestic activities become publicly visible, and which requires frequent shifts in self-presentation to different audience members, as Goffman's (1961) work, discussed previously, helps us to understand.

The Public and the Private Self

The exchange between internalising others' perceptions of us and the formation of self-concept has been systematically studied and discussed in illuminating ways by Tice. In Tice's (1992) seminal paper 'Self-concept change and presentation: The looking glass self is also a magnifying glass', a series of social psychological experiments were presented which aimed to examine the association between the publicness of a person's behaviour and the subsequent impact on their self-concept. Essentially this series of experiments tested the effects of self-presentation on the concept of the self. In the experiments, participants were asked to take on different dispositions and to do so either publicly or privately. Results from this series of experiments concluded that the "self-concept is more likely to change by internalising public behaviour than by internalising identical behaviour but lacks interpersonal context" (p. 435). Indeed, those participants who expected future interaction with an audience were more likely to internalise the self-concept that was being made salient.

What relevance do these findings have for clinical forensic practice? One element of forensic practice where such findings could hold significance is psychological or behavioural group programming. As part of sentence planning, prisoners may be required to engage in 'offending behaviour' and other skills-based programmes (Ministry of Justice, 2021). These programmes range in their focus and include managing anger, cognitive behaviour, social skills training, victim awareness, and sexual offending courses. Some violence reduction programmes, for example, involve structured group sessions (Wong & Gordon, 2013). Such groups tend to focus on changing 'cognitions' in relation to offending. Usually these groups consist of a number of participants at a time with two or three facilitators. The groups are often interactive in nature and require participants to carry out activities or engage in role plays.

While these groups are closed to participants selected for that particular group, there is a 'public' element to it. Responses given by participants are responded to by other participants and by the group facilitators. The developing social norms of the group are of relevance here. It could be argued that the 'audience' of the group might shape the extent to which behaviour is internalised. For example, if someone in the group feels judged for presenting a particular point of view, this may affect what that person chooses to disclose in the future. Likewise, observing someone else receiving group criticism (even if it is constructively framed) or, alternatively, group praise, could also be internalised by other group members and subsequently influence the way they present themselves to the group in the future. As a result, the group setting may, inadvertently, result in an inauthentic space where there is much impression management of participants and little genuine sharing of a person's core or vulnerable self. Maintaining an awareness of these group dynamics is important for group facilitators who aim to create a 'safe space' in which all viewpoints can be shared and are welcomed.

A specific group programme which has a strong interpersonal 'audience' context and is utilised in some forensic settings in the UK, is the Sycamore Tree Programme (Feasey & Williams, 2009; Fourie & Koen, 2018). This programme includes some restorative justice elements in that it is a victim awareness programme. Developed by the Prison Fellowship, a charity working within prison settings, it is a group-based intervention which takes a multi-modal approach to delivery. The groups consist of small and large group work (called 'circles') and a number of interactive tasks. During the programme, victims of offences (but not the specific victims of those taking part in the programme) come in to talk about their experience and engage in a discussion with the participants. During the programme, participants also engage in an 'act of restitution' where they prepare a picture or poem and then present this to the 'community' (to an audience of other staff in the prison or hospital). Harvey and Drennan (2021) conducted a focused-ethnography in one medium secure unit in the UK to explore the implementation of the Sycamore Tree Programme. They found that for facilitators to effectively contain the emotional dimensions of the programme, they needed to take a relational and collaborative approach. Harvey and Drennan make sense of their findings using Goffman's (1959) *The Presentation of Self in Everyday Life*, arguing that it was just as important for the facilitators as it was for the participants to be prepared to bring some of their 'back stage' emotional vulnerabilities to the 'front stage'. Doing so created the opportunity for a more poignant shared experience of their common humanity. It was, however, also important for facilitators to be highly skilled at 'containing' the group by holding their own and the group's boundaries in safe and grounding ways. Nevertheless, Harvey and Drennan argue that being exposed to emotion in the present enabled the group to engage in a meaningful human encounter. These insights hint at the role that practitioners can play in supporting clients to meaningfully move between the front and backstage of their self-presentation by, at times, role-modelling to clients how to safely do this. By skilfully supporting clients to remove te 'mask' and authentically relate to others, practitioners and clients can break down less adaptive self-presentation barriers together. At the same time, practitioners can help support clients to decide when to skilfully use impression management techniques when it is necessary for them to do so to keep themselves safe. We will return to these ideas again later on in the chapter, but considering how more 'authentic' presentations of self can be safely facilitated with clients, it is an important area to explore further.

Related to the ideas above, it is of relevance to consider compassion focused therapy (CFT) as an approach that can decrease distress and improve wellbeing (Gilbert & Procter, 2006) and which, in turn, may help practitioners to create safe-spaces where change is possible. CFT has roots in evolutionary psychology and grew out of a developing understanding of human brain and social development (Gilbert, 2014). Of particular relevance to our concerns in this paper is the recognition within CFT that the human brain is shaped by social processing, and that social relationships influence the ways in which thought processes are organised. In particular, some of the research that informs CFT argues that affectionate and caring relationships result in beneficial effects on human physiology and psychology (Cozolino, 2007; Siegel, 2012). CFT is also informed by research which recognises the importance of the relationship we have with ourselves and, in particular, experiences of shame and self-criticism (Kannan & Levitt, 2013; Kim et al., 2011). Gilbert points out that working with clients to develop their compassion can have a wide range of physiological, psychological, and therapeutic benefits. From a social psychological perspective, it is worth considering how practitioners, when working one-to-one or facilitating group-work with clients, can enable and encourage compassion, kindness, and consideration for others as well as positive self-regard. In the often abrasive or cold environment of prisons and other secure settings, practitioners may need to think carefully about both their own self-presentation as well as those of their clients if they want to foster more compassionate, relational dynamics and/or create spaces where expressions of self-compassion can be safely met. On the basis of the research that informs CFT, it might be argued that a compassionate 'atmosphere' in group work or one-to-one client work in secure settings would allow clients to feel safe enough to 'drop the mask' of impression management so that they are enabled to find and express more of their authentic self.

It is also important to bear in mind that when we consider the social context in which group programmes (or indeed individual work) occurs, there are a number of other competing factors that encourage or sustain various forms of self-presentation. For example, in some prison settings, for some people to feel able to survive, they have to present themselves as 'hard', 'untouchable', and 'fearless'. Recall also the advice that Ryan was given by his cell mate and the induction officer at the start of this chapter about how he needed to present himself. The prison environment needs to be taken account of if the aim of a group is to start to shift a person's self-concept in a more pro-social and compassion-focused direction. That is, it can be important for many aspects of surviving prison life that prisoners maintain a 'mask' because there are certain forms of self-presentation that are needed for psychological, and at times physical, survival. Indeed, any rehabilitation efforts need to be considered in tandem with the psychological processes at play within the environment. Is it ethical for forensic practitioners to encourage prisoners to 'remove' their mask in order to bring about the therapeutic change the client may need or want? If such therapeutic work is carefully done in a prison setting, then there may also be a responsibility on the part of the practitioner to help the prisoner become more skilful at consciously putting their mask back on before going back onto the wing. Perhaps it is the role of the practitioner to help their clients know when and when not to remove their 'masks' in order to navigate competing demands for an 'authentic self' during the process of rehabilitation, versus managing the form of self-presentation that is needed to survive 'doing time'. Before we dive more deeply into considering the complexities around the formation of a client's self-concept, let us first consider some of the conscious motivations that lead people to actively 'impression manage'.

Motivations for Strategic Impression Management Construction

Self-presentation, or impression management, has different underlying motives. Jones and Pittman (1982) argue that the performance of self-presentation occurs when the actor cares about the impression that others form of them. Strategic self-presentation is the aspect of behaviour that is "affected by power augmentation motives designed to elicit or shape others' attributions of the actor's dispositions" (Jones & Pittman, 1982, p. 223). They also argue that a person becomes aware of themselves through the interactions she has with others. This awareness of the way another appears to see us, influences how we see ourselves. Rather than viewing strategic self-presentation as a deceptive process, Jones and Pittman propose that the features of it "typically involve selective disclosures and omissions" (ibid.). They outline a range of potential different motives involved in self-presentation, including: *ingratiation* (seeking to be likeable), *intimidation* (wanting people to think they are dangerous), *self-promotion* (seeking the attribution of competence), *exemplification* (projecting integrity and morality), and *supplication* (when the individual lacks resources to enact other motives, presenting the self as weak and helpless).

Leary and Kowalski (1990) aimed to pare down the number of variables that affect impression management with a view to identify the most salient and theoretically meaningful factors. As part of this process, they identified a two-component discrete process which both operate differently according to differing principles, situations, and dispositions. The first component is *impression motivation* and the second is *impression construction*. They argue that the "impression motivation process is associated with the desire to create particular impressions in others' minds, but may or may not manifest itself in overt impression-relevant actions" (p. 35). The impression construction process activates once a person is motivated to create a particular impression, and involves altering one's behaviours to try and influence the impressions others have of them. They argue that impression construction is influenced by five factors: self-concept, desired or undesired identities, constraints associated with roles the individual may hold, the values of the person they are seeking to manage the impression of, and their own perceptions of how they are currently 'seen' or regarded. Some of these ideas have been further expounded upon by other authors under the banner of self-discrepancy theory, which we examine later in the chapter.

But the recognition that there are different processes and motives behind impression management strategies is important for forensic practitioners. The self-presentation strategies and motivations that clients might use, may be explained by their desire to successfully progress through their sentence or to cope and survive being in the, sometimes hostile, environment of a secure setting. Given the competing demands of prison life, it is possible that a person in prison may shift between attempting to be seen as 'morally responsible' when engaging with staff involved in their rehabilitation, and trying to project an impression of someone to keep your distance from within the prison community. Different individuals will opt for different self-presentation tactics in order to navigate the complex interplay of relationships that secure settings demand. To be seen as 'dangerous' might aid survival. Conversely, being liked or seen as competent may also be advantageous under certain circumstances.

Self-discrepancy Theory and the Experiences of Shame and Guilt

Our decisions on how to manage our impressions externally will depend not only on how we see ourselves but how we imagine others see us too. Higgins's (1987) self-discrepancy theory – a crucial theory that links to impression management – argues that there are competing perceptions of how we see ourselves and how we perceive others see us. Higgins posits that there are three domains: the 'actual' self (the attributes that a person believes they have), the 'ideal' self (the kind of person an individual wishes they were), and the 'ought' self (the kind of person an individual believes they ought to be like as part of their duty). In addition to these three domains, Higgins suggests that there are two standpoints from which the self can be viewed: from the person's *own* standpoint (what the person thinks about themselves) and from a *significant other's* standpoint (what the person thinks a significant other thinks). Combining the domains and the standpoints results in six self-states (Table 4.1).

The actual self-states ('actual-own' and 'actual-other') are referred to as the 'self-concept' and the other four self-states presented in Table 4.1 are referred to as 'self-guides' or 'standards of being' (Higgins et al., 1985; Tangney et al., 1998). It is argued that people are motivated to ensure that there is a match between their 'self-concept' and their 'self-guide'. It is argued that different forms of discrepancy give rise to different emotional reactions. These are presented below (Table 4.2).

Table 4.1 Self-states

Selves	Standpoint	
	Own	Other
Actual	actual-own [1] ('self-state')	actual-other [2] ('self-state')
Ideal	ideal-own [3] ('self-guide')	ideal-other [4] ('self-guide')
Ought	ought-own [5] ('self-guide')	ought-other [6] ('self-guide')

Table 4.2 Discrepancies between self-states and self-guides leading to emotional reactions

Discrepancy	Outcome	Example
'actual-own' (from own or another's standpoint) versus 'ideal-own' [1] versus [3]	A person's attributes differ from what the person wished or hoped for. Through not obtaining the attributes wished or hoped for, the person can feel dejected (disappointed, frustrated, and dissatisfied).	When a person in prison reads a psychological report about themselves that attributes characteristics to them that they do not recognise themselves as having, they can feel frustrated by the sense that they have been misunderstood or seen only in part, but not in the round.
'actual-own' (from own or other standpoint) versus 'ideal-other' [1] versus [4]	A person's attributes differ from the ideal attributes that a significant other wished or hoped for. Through not obtaining the attributes that another wished or hoped for by a significant other, the person can feel dejected and can be vulnerable to experiencing shame and embarrassment.	A prison officer saying that he "wouldn't have expected this from you. I thought you were better than this". Shame and embarrassment might result in this situation.
'actual-own' (from own or other standpoint) versus 'ought-own' [1] versus [5]	A person's actual attributes differ from what the person feels he/she ought to have. Through this discrepancy, the person can experience agitated-related emotions and be vulnerable to feelings of guilt.	When a person is confronted with the idea that they aren't the father or mother they wanted to be to their children. A parent in prison may feel guilt in this situation because they had aimed for the ideal of being a better parent than their own parents were to them.

Discrepancy	Outcome	Example
'actual-own' (from own or other standpoint) versus 'ought-other' [1] versus [6]	A person's actual attributes differ from what a significant other feels the person ought to have. Through this discrepancy, the person can experience agitated-related emotions due to fear of punishment. The person might fear being harmed.	Building on the above example, when a child says to the parent "you weren't there for me as a mother/father". For a person in prison, this revelation by their child may stimulate fear that the child will not want to have further contact with them and they will be further ostracised from the people they love most.

The above table summarises some examples of the way self-discrepancy theory can operate and result in different emotional reactions that also reinforce or increase self-criticism. It should be noted that the different self-discrepancy scenarios result in different emotional experiences. For example, Higgins et al. (1985) argue that when there is a self-discrepancy between the *actual/own* versus the *ideal/other*, the person is prone to experience shame. But if there's a discrepancy between the *actual/own* and the *ought/own*, the person is prone to experience guilt. These findings, however, have been challenged by Tangney et al. (1998) who found that all self-discrepancies would result in shame, but not necessarily guilt. Although there is not a definitive conclusion emergent in the research literature on the range of emotions that self-discrepancy can arouse, the issue of shame is of particular importance in relation to impression management. For example, if a strong shame response arises as a result of a client 'seeing' a discrepancy for the first time, it could, in part, explain how therapy can sometimes be destabilising or iatrogenic for people who have been unaware of, or avoided, connecting with discrepancies.

Golding and Hughes (2012) have aimed to distinguish the internal processes that are associated with guilt and shame, respectively. Guilt, they argue, is associated with one's behaviour. It is an "other-centred emotion that exists in response to the perception that one has hurt another through one's behaviour. Guilt is associated with empathy…" (Golding & Hughes, 2012, p. 94). Shame, by contrast, is argued to be a self-centred emotion that reduces empathy for others. Shame is activated in an effort to protect oneself from attacks from another. Moreover, the experience of shame tends to result in the avoidance of others and the loss of attachment or connection with others. Experiences of shame are more likely to decrease attachment behaviours. Conversely, guilt can motivate the drive to reach out to another person to make amends.

In early childhood development, Golding and Hughes (2012) suggest that shame precedes guilt. When a child is scolded and shame is experienced, the child may disconnect from the parent or guardian and try to hide, freeze, and/or become silent. If a parent or guardian then comforts the child appropriately and helps them to make sense of what has happened, the child can reconnect, and shame reduces. By gently calming the child and making the space between parent and child safe again, the parent can then explain what they did wrong. Through this re-secured bond between parent and child, the child then has the space to think and see what they did wrong for themselves. Golding and Hughes argue that when this is successfully done in early childhood, the child learns "to think about what she did and make sense of it. In this way your child begins to perceive the consequences of her actions, learns to experience empathy for the other and to experience guilt over the other's distress" (p. 156). However, when shame is not skilfully resolved in this way, a person can develop what Golding and Hughes have called a "shield against shame" (p. 156). This consists of defensive behaviours that prevent someone from having to see or experience the reaction of another. Golding and Hughes state that, as a result of the 'shield against shame', the child "denies or minimizes his behaviour, he blames others when things go wrong, and, if all else fails, he rages against himself and others in his world" (p. 155). They argue that in attachment relationships, the three qualities of relationship that need to be present to help children successfully resolve their shame responses include: curiosity, acceptance, and empathy. These insights on the way guilt, shame, and the resolution of shame, work in early childhood development are, arguably, just as true for adult attachment and adult therapeutic relationships. When applied to the field of forensic practice, this may help us to better understand the actions and reactions of people confined in forensic environments, particularly with reference to their self-presentation. It could also provide a fruitful 'way in' to resolving unresolved shame experiences. After all, as Gilligan (2003) has observed through his extensive work in prisons and secure mental health facilities, unresolved shame and guilt play a significant role in the life histories of those confined in these environments, not least due to the link between unresolved shame and violence. We now add to this previous work and suggest that the 'shield against shame' may be part of what the self-presentation mask is made of for many people living in confinement.

Returning to the work of Higgins et al. (1985, 1986) and Higgins (1987) and self-discrepancy theory, what is important for our purposes here in seeking to understand the adaptive and mal-adaptive functions of impression management is to acknowledge that adults differ in relation to which self-guide they are motivated by. People can be motivated to reach a condition where the self-concept (actual self) matches their aspirational self-guide. The implications of this for self-presentation, return us to our previous points about the potential importance of creating a relationally 'safe' atmosphere where vulnerabilities can be expressed, seen, and empathised with. When thinking about impression management in forensic settings, it is of interest to view it from the angle of self-discrepancies and the 'work' the person is doing to avoid (perceived or real) threat or attack in the environment. Recall that shame is often activated in order to avoid having to face the attack, judgement, or retribution of another. Impression management, in this sense, can be a tried and true defence mechanism that will be very difficult for the practitioner to break down. These ideas invite us to question the extent to which forensic practitioners are aware of the discrepancies that exist in their clients, and the way these discrepancies can link to shame. What attempts might be made to narrow the gap of self-discrepancy to allow for a skilfully managed dismantling of the shield of shame? Should we be measuring self-discrepancies in routine clinical practice as a first step in assessing the depth of disconnection a person feels in relation to their own sense of self and/or in relation to others? Could this be a means of helping clients to begin to resolve unprocessed shame? What does the criminal justice system activate for people in relation to self-discrepancies? Following what we have noted above from the work of Golding and Hughes (2012) about the difference between shame and guilt, how can forensic practitioners better facilitate therapeutic encounters that draw out and re-process buried experiences of shame? If shame is an instinctive natural response to our own wrongdoing, then, as an emotional experience, it needs to be carefully met and processed in order for it to be transformed into a healthy version of guilt that, in turn, can be the driver that leads a person to want to prosocially reconnect, make amends, and form (or re-establish) successful attachment and social bonds. To do this kind of work in a forensic setting, it would be important for practitioners to tread carefully when trying to work with clients to remove their shield of shame and to focus on creating a relational field with the client that allows for curiosity, acceptance, and empathy.

Conclusion: The Forensic Practitioner's Impression Management

Throughout the chapter, we have mostly focused on impression management from the standpoint of the client. However, we have also, at times, referred to the fact that all social interactions require some degree of impression management for all of us. Likewise, staff working in forensic settings do not exist in a vacuum – they too are subject to different interpersonal challenges of working in forensic settings. Like everyone else, forensic practitioners will engage in a process of impression management towards other professionals and towards their clients. The practitioner might be keen to ensure that they can establish engagement with a client and be consciously thinking about how they are 'coming across' to their client: Am I being sufficiently boundaried? Am I listening enough and coming across as empathic? Have I made it clear to the client the limits of confidentiality? It is important that the presentation of the self is managed in a careful way to ensure engagement and that the client feels safe in the process. It is also important for practitioners to be authentic and to role model ways of sharing their authentic self, dropping their own mask at times and finding ways of safely making connections with clients on the basis of their shared human vulnerabilities. For those working in non-forensic settings, it might be much easier to role model to clients. In some therapeutic techniques, self-disclosure is framed as part of the limited re-parenting role of the therapists (Rafaeli et al., 2011). However, such a technique would likely be deemed as inappropriate in most forensic settings. It is, of course, important to remain mindful of the context in which forensic practitioners work, ensuring that their practice remains consistent with security protocols. But authenticity can be approximately role modelled in simple ways by sharing carefully chosen examples which demonstrate that, at times, everyone worries about what others think, and engages in some form of impression management. Forensic practitioners could conversationally offer to clients a range of vignettes (not taken from their own personal lives) that fit with the circumstances the client has found themselves in, as a means of demonstrating that they are not alone, and that all human beings struggle with such challenges. Some tips for practice are presented (Box 4.1).

Box 4.1 Taking an impression management lens in practice

- Reflect on how your own impression management might impact upon your work with the client and their response to you.
- Reflect on how clients might manage their impression as an adaptive response to a closed institution. How might this be considered within psychological formulations?
- In group settings, people might be prone to experiencing shame that might heighten the need to manage their impressions towards others. Trust and psychological safety are prerequisites to allow group participants to disclose and show emotions.
- A cohesive multi-disciplinary team is a prerequisite for practitioners to be able to authentically reveal the self in a boundaried manner.
- It is important to reflect on the role of shame in relation to impression management. Shame can drive presentation and can also occur when a client sees a discrepancy in their self-concept. It is important for practitioners to be aware of this and that the client might have a psychological response to shame as a result.

In essence, we argue that acknowledging the context of secure settings must always remain central to the interactional space and to ensuring people keep whatever boundaries they need to remain safe. An ever-present awareness of the social environment in which we are in, is at the heart of understanding the way people 'self-manage', including the ways their presentation of self is constructed, perceived, and distorted. Moreover, acknowledging context allows for recognition and understanding of the ways in which different aspects of the self are rendered, revealed, or concealed, according to the interplays of relational dynamics, self-preservation techniques, and previous attachment-related experiences. We suggest that applying a social psychological lens aids the development of a deeper and more nuanced understanding of impression management, enabling a psychological understanding and formulation of both the client's and practitioner's interactional style. Such understanding then helps us to make sense of our own behaviour and the behaviour of others, and the approaches we take to self-presentation and impression management in complex, high stakes, social environments.

References

Book, A. S., Holden, R. R., Starzyk, K. B., Wasylkiw, L., & Edwards, M. J. (2006). Psychopathic traits and experimentally induced deception in self-report assessment. *Personality and Individual Differences*, 41(4), 601–608.

Bowlby, J. (1969). *Attachment and loss*. Basic Books.

Cozolino, L. (2007). *The neuroscience of human relationships: Attachment and the developing brain*. Norton.

Douglas, K. S., & Reeves, K. A. (2010). Historical-Clinical-Risk Management-20 (HCR-20) violence risk assessment scheme: Rationale, application, and empirical overview. In R. K. Otto & K. S. Douglas (eds), *International perspectives on forensic mental health. Handbook of violence risk assessment* (pp. 147–185). Routledge.

Drake, D. (2006). A comparison of quality of life, order and legitimacy in two maximum-security prisons. Unpublished PhD thesis, University of Cambridge.

Edens, J. F., Buffington, J. K., Tomicic, T. L., & Riley, B. D. (2001). Effects of positive impression management on the Psychopathic Personality Inventory. *Law and Human Behavior*, 25, 235–256. doi:10.1023/A:1010793810896.

Edens, J. F., & Ruiz, M. A. (2006). On the validity of validity scales: The importance of defensive responding in the prediction of institutional misconduct. *Psychological Assessment*, 18, 220–224. doi:10.1037/1040-3590.18.2.220.

Emery, L. G., Muise, A., Dix, E. L., & Le. B. (2014). Can you tell that I'm in a relationship? Attachment and relationship visibility on Facebook. *Personality and Social Psychology Bulletin*, 40(11), 1466–1479.

Feasey, S., & Williams, P. (2009). An evaluation of the Sycamore Tree programme: based on an analysis of Crime Pics II data. Retrieved from http://shura.shu.ac.uk/1000/1/fulltext.pdf.

Flynn, S., Noone, C., & Sarma, K.M. (2018). An exploration of the link between adult attachment and problematic Facebook use. *BMC Psychology* 6(1), 34.

Fourie, M. E., & Koen, V. (2018). South African female offenders' experiences of the Sycamore Tree Project with strength-based activities. *International Journal of Restorative Justice*, 1(1), 57–80. doi:10.5553/IJRJ/258908912018001001004.

Gilbert, P. (2014). The origins and nature of compassion focused therapy. *British Journal of Clinical Psychology*, 53, 6–41.

Gilbert, P., & Procter, S. (2006). Compassionate mind training for people with high shame and self-criticism: Overview and pilot study of a group therapy approach. *Clinical Psychology & Psychotherapy: An International Journal of Theory & Practice*, 13 (6), 353–379.

Gillard, N. D., & Rogers, R. (2015). Denial of risk: The effects of positive impression management on risk assessments for psychopathic and nonpsychopathic offenders. *International Journal of Law and Psychiatry*, 42–43, 106–113. doi:10.1016/j.ijlp.2015.08.014.

Gilligan, J. (2003). Shame, guilt and violence. *Social Research*, 70(4), 1149–1180.

Goffman, E. (1959). *The presentation of self in everyday life*. Penguin.
Goffman, E. (1961). *Asylums*. Doubleday.
Golding, K., & Hughes, D. (2012). *Creating loving attachments: Parenting with PACE to nurture confidence and security in the troubled child*. Jessica Kingsley Publishers.
Gozna, L.F., Vrij, A., & Bull, R. (2001). The impact of individual differences on perceptions of lying in everyday life and in a high stake situation. *Personality and Individual Differences*, 31(7), 1203–1216.
Hare, R. D. (1998). The Hare PCL-R: Some issues concerning its use and misuse. *Legal and criminological psychology*, 3(1), 99–119.
Harvey, J., & Drennan, G. (2021). Performing restorative justice: Facilitator experience of delivery of the Sycamore Tree Programme in a forensic mental health unit. *International Journal of Restorative Justice*.
Harvey, J. (2012). *Young men in prison: Surviving and adapting to life inside*. Routledge.
Higgins, E. T. (1987). Self-discrepancy: A theory relating self and affect. *Psychological Review*, 94, 319–340.
Higgins, E. T, Klein, R., & Strauman, T. (1985). Self-concept discrepancy theory: A psychological model for distinguishing among different aspects of depression and anxiety. *Social Cognition*, 3, 51–76.
Higgins, E.T., Bond, R. N., Klein, R., & Strauman, T. (1986). Self-discrepancies and emotional vulnerability: How magnitude, accessibility, and type of discrepancy influence affect. *Journal of Personality and Social Psychology*, 51, 5–15.
Jones, E. E., & Pittman, T S. (1982). Toward a general theory of strategic self-presentation. In J. Suls (ed.), *Psychological perspectives of the self* (vol. 1, pp. 31–261). Erlbaum.
Kannan, D., & Levitt, H. M. (2013). A review of client self-criticism in psychotherapy. *Journal of Psychotherapy Integration*, 23, 166–178. doi:10.1037/a0032355.
Kim, S., Thibodeau, R., & Jorgensen, R. S. (2011). Shame, guilt, and depressive symptoms: A meta-analytic review. *Psychological Bulletin*, 137, 68–96. doi:10.1037/a0021466.
Knight, K., Simpson, D. D., & Hiller, M. L. (2002). Screening and referral for substance-abuse treatment in the criminal justice system. In C. G. Leukefeld, F. M. Tims, & D. Farabee (eds), *Treatment of drug offenders: Policies and issues* (pp. 259–272). Springer.
Leary, M. R. (2019). *Self-presentation: Impression management and interpersonal behavior*. Routledge.
Leary, M. R. & Kowalski, R. H. (1990). Impression management: A literature review and two-component model. *Psychological Bulletin*, 107(I), 34–47.
McEwan, T. E., Davis, M. R., MacKenzie, R., & Mullen, P. E. (2009). The effects of social desirability response bias on STAXI-2 profiles in a clinical forensic sample. *British Journal of Clinical Psychology*, 48, 431–436. doi:10.1348/014466509X454886.
Mead, G. H. (1934). *Mind, self, and society from the standpoint of a social behaviorist*. University of Chicago Press.
Mills, J. F., & Kroner, D.G. (2006). Impression management and self-report among violent offenders. *Journal of Interpersonal Violence*, 21(2), 178–192.
Ministry of Justice. (2021). Offending behavioural programmes. Retrieved from https://s3.amazonaws.com/thegovernmentsays-files/content/169/1697081.html.
Nathanson, D. L. (ed.). (1987). *The many faces of shame*. Guilford Press.
Ort, D. (2020). Opening up while locked down: Client disclosure in correctional settings. *Journal of Clinical Psychology*, 76, 308–321.
Paulhus, D. L. (1986). Self-deception and impression management in test responses. In A. Angleitner & J. S. Wiggins (eds), *Personality assessment via questionnaire* (pp. 143–165). Springer-Verlag.

Paulhus, D. L. (1998). *Paulhus Deception Scales (PDS): The Balanced Inventory of Desirable Responding-7: User's manual*. Multi-Health Systems.

Putnam, R. D. (1995). Bowling alone: America's declining social capital. *Journal of Democracy*, 6, 65–78.

Rafaeli, E., Bernstein, D. P., & Young, J. (2011). *Schema therapy*. Routledge.

Rogers, R. E., & Bender, S.D. (2018). *Clinical assessment of malingering and deception*. Guilford Press.

Roth, D. L., Harris, R. N., & Snyder, C. R. (1988). An individual differences measure of attributive and repudiative tactics of favorable self-presentation. *Journal of Social and Clinical Psychology*, 6(2), 159–170.

Siegel, D. (2012). *The developing mind: How relationships and the brain interact to shape who we are* (2nd edition). Guilford Press.

Simms-Sawyers, C. Miles, H. & Harvey, J. (2020) An exploration of perceived coercion into psychological assessment and treatment within a low secure forensic mental health service. *Psychiatry, Psychology and Law*, 27(4), 578–600. doi:10.1080/13218719.2020.1734981.

Tangney, J. P., Niedenthal, P. M., Covert, M. V., & Barlow, D. H. (1998). Are shame and guilt related to distinct self-discrepancies? A test of Higgins's (1987) hypotheses. *Journal of Personality and Social Psychology*, 75(1), 256.

Tice, D. M. (1992). Self-concept change and self-presentation: The looking glass self is also a magnifying glass. *Journal of Personality and Social Psychology*, 63(3), 435.

Tully, R. & Bailey, T. (2017). Validation of the Paulhus Deception Scales (PDS) in the UK and examination of the links between PDS and personality. *Journal of Criminological Research, Policy and Practice*, 3, 38–50. doi:10.1108/JCRPP-10-2016-0027.

Ugelvik, T. (2016). *'Be a man. Not a bitch.' Snitching, the inmate code and the narrative reconstruction of masculinity in a Norwegian prison*. Ashgate.

Uziel, L. (2010). Rethinking social desirability scales: From impression management to interpersonally oriented self-control. *Perspectives on Psychological Science*, 5, 243–262. doi:10.1177/1745691610369465.

Walters, G. D. (1995). The Psychological Inventory of Criminal Thinking Styles: Part I. Reliability and preliminary validity. *Criminal Justice and Behavior*, 22, 307–325.

Williams, M. M., Rogers, R., & Hartigan, S. E. (2020). The validity of the PICTS-SV and its effectiveness with positive impression management: An investigation in a court-mandated substance use treatment facility. *Criminal Justice and Behavior*, 47(1), 80–98. doi:10.1177/0093854819879733.

Wong, S. C., & Gordon, A. (2013). The violence reduction programme: A treatment programme for violence-prone forensic clients. *Psychology, Crime & Law*, 19(5–6), 461–475.

5 Attitudes and Beliefs

Lara Arsuffi

This chapter focuses on applying social psychology research on "attitudes and beliefs" to the assessment and rehabilitation of people in forensic settings, who generally present with a range of complex difficulties, including an offending history, substance misuse issues, a complex trauma history, and difficulties in developing trusting relationships with professionals involved in their care. This chapter will first explore some of the concepts which help to explain attitudes and their formation, and will describe how these concepts have been researched in social psychological studies and, subsequently, applied to forensic practice. It will examine the importance of *social learning* and the *mere exposure effect* (Zajonc, 1968) and consider how people exposed to higher rates of abuse and neglect in their developmental years may end up offending due to their attitudes towards violence and anti-social behaviour. It will consider the role of *schemas* (Young et al., 2003) and their application to personality theory, and how practitioners consider the role of schemas when working with clients who have been diagnosed with personality difficulties. In addition, this chapter will consider how *implicit* and *explicit attitudes* might be important for delivering intervention, and consider the role that *cognitive dissonance* (Festinger,1957) might play in achieving change.

The Formation of Attitudes and their Role in the Evaluation of the Social World

Attitudes, in social psychology, refer to the cognitions that people hold about themselves, objects, events, or other people. Attitudes are thought to serve the function of facilitating the adaptation of a person to their environment, and have been shown to substantially influence behaviour (Fazio & Olson, 2003). For example, positive attitudes towards violence and aggression have been found to predict a diverse range of aggressive behaviours, such as bullying at school (Eliot & Cornell, 2009), domestic violence in college students (Fincham et al., 2008), domestic violence in adult men attending intimate partner violence interventions (Eckhardt et al., 2012), and teachers' use of violence towards their students (Khoury-Kassabri, 2012).

DOI: 10.4324/9781315560243-5

The literature describes two types of attitudes: implicit attitudes and explicit attitudes. Implicit attitudes (IA) have been defined as "introspectively unidentified traces of past experiences that mediate ... feelings, thoughts, or action towards a social object" (Greenwald & Banaji, 1995, p. 8). IAs are thus automatic affective reactions activated in the face of a relevant stimulus (Gawronksi et al., 2003), which are manifested in actions or judgements without the individual's conscious awareness or control. By contrast, explicit attitudes (EAs) are defined as conscious attitudes that a person chooses to display while interacting with the world (Rydell et al., 2008). For instance, a person may have grown up in a family or culture where extreme prejudice against other ethnic groups (implicit attitudes) are deeply ingrained. However, with education and positive interactions with other ethnic groups, over time, a person may choose to adopt an inclusive and non-prejudiced "explicit attitude" towards people from other ethnic groups.

Andrews and Bonta's (2010) Risk-Need-Responsivity Model (RNR) of forensic populations' rehabilitation highlights the importance of addressing explicit crime-supportive attitudes (which are considered dynamic risk factors) through intervention programmes as meta-analyses have consistently found that they are related to recidivism (Andrews & Bonta, 2010; Helmus et al., 2013).

Attitudes and the Prediction of Aggression

In aggression research EAs are usually evaluated using self-report measures (i.e. interviews or psychometric questionnaires). Andrews and Bonta (2010) have classified explicit procriminal attitudes (PCA) measures in three categories. These are (1) identification with criminal others, (2) rejection of convention, and (3) techniques of neutralisation. Identification with criminal others refers to the approval and imitation of antisocial models; rejection of convention includes beliefs that stress opposition to the legal system and people working within it (e.g. police), and techniques of neutralisation relate to justification of criminal behaviour (Banse et al., 2013). These distorted cognitions have been argued to be one of the most notable characteristics of forensic populations.

Banse et al. (2013) suggest that, alongside exploring the impact of explicit PCAs on risk of offending, future research should investigate the impact of implicit attitudes (IAs) on criminal behaviour as some recent studies have provided empirical evidence that IAs also strongly affect behaviour. IAs are measured indirectly through means such as priming and the Implicit Association Test (IAT; see Fazio & Olson, 2003, for a review). These indirect measures do not rely on self-report and therefore are less likely to be affected by social desirability biases (see Chapter 4). Though not true for all measures, it is often the case that the participant is unaware that attitudes and stereotypes are being assessed (Fazio & Olson, 2003).

For example, in a classic study by Higgins et al. (1977), participants were asked to practice words while performing a perception task. Some of the words were trait concepts such as 'brave' or 'careless'. Then, in the context of a reading comprehension study, participants were presented with a description of a person's behaviour that was unclear in relation to the primed dimension. For example, the person was said to enjoy being a risk-taker. After reading this description, the participants were asked to rate how much they liked the person. Participants who were shown positive traits related to the behaviours (e.g. 'brave') liked the target person more than those who were exposed to negative descriptors (e.g. 'careless'). These findings were replicated by other studies (e.g. Bargh & Pietromarco, 1982).

The rationale of these experiments is that priming stimuli in the environment initiate mental representations pertinent to them. As this activation persists, the likelihood of its use in subsequent situations increases (Higgins, 1996). But can these laboratory-based findings translate to real social situations? For example, imagine seeing a violent movie and then, a short while later, witnessing someone pushing a person in the street. How would this pushing behaviour be interpreted? Carver et al. (1983) examined this situation. They showed that watching a short film illustrating an intimidating interaction, impacted how participants later construed the behaviour of another person in an unrelated situation. In this experiment, participants who watched the violent interaction perceived an unclear behaviour as being more intimidating than those who had previously watched an innocuous interaction.

Similarly, on a forensic mental health ward, a group of patients may observe a fellow service user(s) assaulting members of staff or other patients. The response team is called and the patient is then restrained and walked to the seclusion area. Witnessing these events could negatively affect their perception of the ward environment and subsequent interpretations of ambiguous interactions with staff or other clients, especially if exposure to repeated and prolonged violence during childhood predisposed them to develop similarly negative attitudes towards this construct. These attitudes about violence could be intensified further in the context of a forensic ward, considering the characteristics of the client's mix.

Researchers have focused their attention on how implicit attitudes are formed, operate, or change (Fazio & Olson, 2003; Dodge & Crick, 1990; Schneider & Fisk, 1982). Literature highlights the importance of repeated matching of stimuli to responses (Schneider & Fisk, 1982; Schneider & Shiffrin, 1977). Through this repeated process, associations are built in the memory system that come to reflect the regularities of one's life. Priming research shows that indirect experimental manipulations can increase activation of aggressive concepts in memory, that can, in turn, affect people's perceptions and judgements. But can repeated exposure to aggressive cues lead to the development of mental structures which would spontaneously impact the interpretation of new hostile events? Dodge's research (Dodge, 1980; Dodge & Tomlin, 1987; Dodge & Crick, 1990) is especially significant in this regard. In these experiments, Dodge demonstrated that children engaging in abnormally aggressive behaviour tend to see hostile intentions in others' behaviours when these acts are unclear. For example, in one study (Dodge, 1980), children exhibiting aggressive behaviour and children not presenting with aggressive behaviour were shown descriptions of confrontational situations, such as "another child spilling food or a drink on your back". Children who had previously exhibited antagonistic behaviour were 50 per cent more likely to perceive hostile intents than those who had not.

Zelli et al. (1995, 1996) worked with participants who either had self-reported aggression (e.g. they had reported hitting another person in the past year) or did not report being aggressive. Sentences which could have been interpreted in two different ways were then shown to participants (e.g. "Mark was pushed by the policeman out of the way"). This can be interpreted as suggesting concern for Mark's welbeing/safety, or as a sentence indicating hostility. Participants were then provided with word cues and asked to recall the original description. Participants, who had self-reported having been aggressive in the past year, were more likely to recall sentences that included cues related to aggression. In other words, participants with hostility-prone mental structures had interpreted the behaviours as motivated by aggression.

The above studies demonstrate how construct accessibility affects perception and judgement. But do concepts, activated in the course of social perception, have direct and automatic effects on behaviour, as well as perceptions and judgements? Carver et al. (1983) told their participants that they would be teaching another participant (a confederate) by administering rewards for correct answers and punishments for incorrect answers (i.e. electric shocks of varying intensity). Carver et al. (1983) found that participants who had been exposed to aggressive concepts in the priming tasks, administered greater electric shocks than control participants. Research has provided additional evidence that if people consistently and frequently pursue the same goal in the same situation, it is possible that relevant situational features can eventually come to trigger the specific goal-directed behaviour. For example, Bargh et al. (1995) tested the hypothesis that there is an automatic association between power and sex for participants who report inclinations to sexually harass. They found that participants with high scores on scales measuring attractiveness of sexual aggression, pronounced sex-related words more quickly when these were preceded by power words, compared to neutral words. This experiment provided some evidence that, for people who are inclined to sexually harass, the accessibility of power-related concepts automatically leads to increased accessibility of sex-related concepts.

This research on automatic triggering of goals has important implications for aggression. For example, imagine that a person has observed on many occasions that the "*normal*" way of resolving interpersonal problems is by using aggression. If this person has used this strategy and achieved his /her goals, then he/she can develop the belief that harming people, who are perceived as hostile or are seen as blocking one's goals, is the right way to overcome these interpersonal difficulties. In other words, if a child is exposed to prolonged and repeated violence when growing up, their development of explicit and implicit attitudes towards this construct can be affected. For example, it may enhance the development of chronic accessibility of aggressive constructs which later affect how the social environment is interpreted (Todorov & Bargh, 2002).

In summary, implicit and explicit attitudes are automatic processes that form through consistent and repeated association of stimuli to responses. Through this repeated process, associations are built in the memory system that come to reflect the regularities of one's life. Attitudes are useful because they help people to maintain consistency in their lives, allowing people to focus their cognitive resources on new and complex situations. However, attitudes can have their downsides. If people are not aware of the underlying motives for their actions, this lack of awareness may make it difficult to control or change behaviour. In these situations, some people may not recognise their unconscious biases and that their actions can be harmful to others. Within this lack of insight, they then tend to use rationalisations and attribute their behaviour to motives that fit their own understanding of what triggered their actions (Wilson & Brekke, 1994). In forensic practice, it is important to assess and understand forensic populations' attitudes because they are considered to be a dynamic risk factor, which has been found to be related to recidivism, that can be reduced during treatment programmes.

Offending Behaviour Treatment Programmes

Considerable research has explored which are the most efficacious treatments to change or reduce forensic populations' maladaptive attitudes. The cognitive-behavioural therapy (CBT) approach is based on the assumption that maladaptive thinking processes, characteristic of forensic populations, are learned rather than inherent (Andrews & Bonta, 2010). It typically explains anti-social behaviour in terms of various socio-cognitive 'deficits' that significantly impair the capacity to reason, and influences how the individual sees and understands the self, other people, and the world more generally (Ross & Fabiano, 1985). In other words, people who have offended are seen as lacking the social problem-solving skills that are necessary to identify and deal with problems of everyday living (McMurran et al., 2001). The focus of intervention is on changing these maladaptive distorted cognitions. Programmes are typically delivered in small group settings. Participants are encouraged to explore the thinking processes that preceded their criminal behaviour, to self-monitor their thinking, and to identify and 'correct' biased, risky, or deficient, thinking patterns.

However, these cognitions may not have been deficits at all when the child was growing up. In fact, they may have served the purpose of helping the child to survive adversity in a specific, unfavourable, context. As the person grows up, these cognitions are no longer adaptive, and may cause the person to engage in risky behaviours that cause them more problems, rather than helping them to achieve a desired outcome. Therefore, seeing these cognitions primarily as deficits, rather than survival or goal-directed strategies, may hinder a person's motivation to engage in CBT-based rehabilitation interventions, and their responsiveness to them in relation to reducing their future maladaptive behavioural choices. In fact, a recent review (Banse et al., 2013) of literature on explicit PCAs as a causal factor of recidivism, concludes that offending behaviour treatment programmes tend to reduce PCAs in general, but, because of limitations in the research designs of the available literature (e.g. lack of comparable treatment and control groups), further research with more rigorous designs is needed to provide more conclusive evidence on the relation of PCAs, PCA treatment, and recidivism (Banse et al., 2013).

The research reviewed so far in this chapter suggests that attitudes play a key role in the occurrence and maintenance of anti-social behaviours. Long-established attitudes are difficult to modify and require a long period of time and extensive practice to change. Literature has supported this view. For example, Stewart et al. (2010) demonstrated that hundreds of trials of overt rejection of a stereotypic association produced change in later stereotypic activation. Rudman et al. (2001) observed changes regarding prejudice (on an IAT measure) and stereotyping (on a priming measure) among students enrolled in a semester-long prejudice and conflict seminar. DeBono (1987) found that attitudes among participants high on a self-monitor scale,[1] altered more if participants were told they were in disagreement with their peer group. By contrast, people low on the self-monitor scale were less likely to change their attitudes. DeBono & Edmonds (1989) found that individuals low on the self-monitor scale were more likely to change their attitudes when they were acting against their beliefs and values. In summary, with appropriate environmental pairings and peer pressure (when appropriate), well-established attitudes that have been learned in the past have the potential to change.

Based on this research, it is suggested that clinicians working in forensic settings should use explicit and implicit measures to more thoroughly assess the full range of attitudes and beliefs associated with their client's anti-social behaviours and functions. Additionally, clinicians should design intervention plans that emphasise the modification of belief structures that perpetuate negative behavioural patterns, and support clients with achieving their goals in more pro-social manners. Attempts to change attitudes could, for example, incorporate learnings from Zajonc (1968) studies (described later in the chapter); that is, repeated exposure to the preferred attitude should occur inconspicuously throughout interventions so that clients develop more favourable attitudes towards the preferred behaviour. This inconspicuous repeated exposure to more favourable attitudes is achieved by scientifically structuring the environment – alternatively called therapeutic milieu – around operant conditioning principles (Skinner, 1979; described later on in this chapter) to promote behavioural changes, and improve the psychological health and interpersonal functioning of the individual.

Within this therapeutic milieu, every interaction between staff and patients throughout the day is an opportunity for a therapeutic intervention, whereby adaptive behaviours are immediately reinforced through positive feedback, everyone is expected to take responsibility for their behaviour, peer feedback is used as a powerful tool for behavioural change, and inappropriate behaviours are mostly dealt with by the withdrawal of reinforcers, rather than with punishment and/or restrictions (Mahoney et al., 2009). A programme designed to reduce risk behaviours in populations with a history of offending, is the Violence Reduction Programme (VRP; Wong & Gordon, 2013) which argues that treatment should be 24 hours a day, seven days a week. Initially designed in Canadian prisons, the VRP strongly advocates the need to work collaboratively with participants to identify and mitigate criminogenic factors, such as antisocial attitudes, substance misuse, and peer influence, while enhancing their strengths by acquiring pro-social interpersonal, cognitive, and vocational skills that are effective in reducing the likelihood of future violence. The VRP emphasises the importance of social influence in the environment of forensic populations, meaning that all staff (e.g. psychiatric nurses, psychologists, psychiatrists, probation and prison officers, etc.) are potential agents for change through modelling prosocial behaviours and supporting skill generalisation to everyday situations. Evaluations of the VRP indicate that programme participation is linked to the reduction in general offending and risk of violence (Wong & Gordon, 2013; Horgan, Charteris, & Ambrose, 2019). In the UK, a similar approach is used in prison-based therapeutic communities (e.g. Grendon prison). In these environments, peer influence (staff and peers) is a key agent of change. By being part of a prosocial community, participants are supported to develop self-examination, learn social norms, develop effective social skills, and build solidarity with others through role rehearsal.

However, there are also downsides to these environments. As they generally house highly complex individuals, participants may be exposed to peers who continue to hold overt, anti-social attitudes and behave impulsively. Some people, who grew up in an environment where there were no boundaries, may struggle to settle into a therapeutic environment where there are strict social norms where behaviour is controlled by their peers. In addition, as these are high-intensity programmes, they are long term (compared to other interventions) and, consequently, require high levels of motivation from participants to engage fully in the therapeutic milieu. As such, therapeutic communities have high drop-out rates, particularly in the first few months of treatment (Alshomrani, 2020). Furthermore, for long term improvements to occur and be maintained in other settings, all participants (staff and clients) need to maintain high levels of programme integrity (i.e. consistently reinforcing the desired attitudes and behaviours). This is not an easy task in the context of the current economic climate, which has affected staff recruitment, training and retention across different forensic settings.

Box 5.1 Top tips for effective therapeutic environments

- Provide a safe environment with consistent limits.
- Reward positive behaviour through immediate positive feedback.
- Encourage responsibility-taking.
- Encourage group decision-making.
- Enhance personal choice.
- Encourage perspective-taking.
- Validate feelings.
- Maintain communication.
- Encourage social interaction and the development of healthy relationships.
- Offer various interventions that improve emotional, social, and occupational functioning, and improve self-esteem and self-efficacy.
- Build compassion (towards self and others), tolerance, and acceptance.

Social Learning Theory, the Mere Exposure Effect, and the Development of Offending Behaviour

Social learning theory was first applied by Sutherland (1947) to explain the development of (white collar) criminal behaviour. Sutherland argued that crime, like any behaviour, is learned in interaction with other persons. This learning also includes motives and drives (i.e. which needs the person is trying to meet via engagement in the criminal behaviour) and rationalisations and attitudes towards the behaviour, as favourable or unfavourable. In other words, the likelihood of an individual engaging in an unlawful act will increase when the person perceives the act as favourable. But how do people develop more favourable attitudes towards offending acts? Through a series of experiments involving exposure to repeated word lists, Zajonc (1968) was the first to show that people tend to develop a preference for what is familiar to them. In his experiments, he showed that participants rated words, which had been presented to them, more positively than words which had not been presented to them. Zajonc named tthis phenomenon the mere exposure effect and suggested that this takes place unconsciously. With regard to how the mere exposure effect relates to offending behaviour, one can hypothesise that the more a person sees someone close to them using anti-social behaviour to successfully achieve their goals, the more likely this person is to develop favourable attitudes towards the offending behaviour.

Sutherland's (1947) theory was later modified by Burgess and Akers (1966) to incorporate operant theory principles (Skinner, 1953) and Bandura's (1969) work on the modelling of behaviour. The four main concepts of social learning theory have been subsequently summarised by Akers (1998) as follows: (1) differential association; (2) definitions; (3) differential reinforcement; and (4) imitation. These are described as follows: (1) *differential association* emphasises the importance of social relationships between people in their social groups (e.g. friends, families, teachers, etc. and also what they see in the media, through television, internet, etc.). These interactions provide a framework within which social behaviour is learnt; (2) *definitions* refers to the attitudes a person attaches to certain behaviours. In other words, definitions refer to rationalisations, justifications, and/or labels that the person assigns to a behaviour as wrong or right (e.g. "it's fun to run away from the police", "it's OK to steal a car if the owner has left the keys in it", "drugs are fun and not harmful"). Most importantly, within social learning theory, definitions are learned through social reinforcement mechanisms (Nicholson & Higgins, 2017).

The next concept, *differential reinforcement*, refers to the cost–benefit calculation of the anticipated or actual rewards or punishments that result from the behaviour. That is, people are more likely to engage in a behaviour based on certain desirable results involving rewards or punishment. Differential reinforcement was introduced by Skinner's (1953) operant conditioning model, which proposes that behaviour can be reinforced (increased) and/or punished (decreased) through respectively positive and negative reinforcement, and positive and negative punishment. The greater and the more frequent the rewards (e.g. social status, money, or excitement), the greater the likelihood that the behaviour will be positively reinforced. This relationship also occurs when a criminal act is not identified by the authorities and therefore is not followed up by adequate punishment (that is, the behaviour is negatively reinforced). With regards to punishment, behaviour can be discouraged through negative or positive punishment. The latter means applying unwanted consequences to the behaviour, whereas the former means the removal of something valued by the person as a consequence of their behaviour (Nicholson & Higgins, 2017). Finally, *imitation* is modelling a behaviour that was observed in others. This element is related to Bandura's (1979) perspective of vicarious reinforcement in which individuals observe the behaviours modelled by others and also the consequences that follow others' behaviour. For example, an individual may observe an unlawful behaviour committed by another, and see the rewards the person has received, or the lack of punishment that followed their behaviour. As a result they may feel encouraged to repeat the same act through imitation. Imitation can result from observation of people in their social circle, or through observation or their role-models in the media (Nicholson & Higgins, 2017).

Over many years, social learning theory has been used to understand the development of offending behaviour in children and adolescents. For example, links between living in a partner-violent home and subsequent aggressive and antisocial behaviour, are suggested by the '*cycle of violence*' hypothesis (Widom, 1989b) derived from social learning theory. This hypothesis proposes that exposure to violence teaches children that controlling others through coercion and violence is normal and acceptable, and indeed using such strategies helps people to reach their goals. Ireland and Smith (2009) tested this hypothesis and found a significant relationship between exposure to parental violence and adolescent conduct problems. Other studies have looked at other dimensions of family violence. For example, maltreatment has gathered solid status as a predictor of crime and anti-social behaviour during adolescence (Stouthamer-Loeber et al., 2001; Widom, 1989a) and adulthood (Smith et al., 2005; Widom & Maxfield, 2001). Many other studies have tested the cycle of violence hypothesis using various methods (e.g. retrospective or longitudinal studies) and measuring different outcomes (e.g. adolescent or early adulthood outcomes, etc.) For a more comprehensive review, please see Ireland and Smith (2009). Despite conflicting results (depending on what outcomes measures were used, what population was being studied, and what variables were / were not controlled for), the suggestion from social learning theory that patterns of observed and experienced violent interactions may be learned in the home, school, or community, and that these may be later expressed in antisocial interactions in other contexts, seems to be warranted.

Therefore, alongside using implicit and explicit measures to assess the attitudes and beliefs of forensic populations, it is also important to collect a thorough account of their psychosocial history, including information about their childhood, education, employment, past and current relationships (intimate, family, and friends), and history of trauma, as this information will be useful to understand the processes which reinforced and maintained their offending behaviour. In an ideal world with a lot of resources, these thorough assessments would take place at the point of access into a forensic environment (be it a secure hospital, prison, or probation service) and be used to formulate treatment plans that target a person's risk factors, while building on their strengths and protective factors. This is the model that Offender Personality Disorder (OPD) services have tried to adopt with consultations and formulations offered to Probation Practitioners by Clinical or Forensic Psychologists for all service users that are screened into the OPD pathway (HMPPS & NHS England, 2020). In addition, in an ideal world, forensic populations should have access to occupational therapy activities, such as employment and education, sport, cooking and other vocational activities, to develop prosocial goals, build their skills/ confidence, and engage positively with others. All staff should have a thorough understanding of operant conditioning and social learning principles, and be skilled in applying these consistently to shape behaviour. Alongside reinforcing adaptive behaviours and weakening unhelpful behaviours, therapeutic interventions could be optimised by incorporating learnings from research on cognitive dissonance (discussed in the next section of this chapter). The use of visual images and technology could be incorporated in treatment environments to target implicit cognitions by repeatedly and subtly exposing participants to more adaptive attitudes. Advertisements in the media use these principles to sell their products, but the ethics of this approach and its application to patient's treatment have not yet been researched. Arguably, many improvements are urgently needed in clinical practice with forensic populations. For example, forensic hospitals tend to use therapeutic milieu principles, but their ability to implement strategies consistently is hampered by high staff turnover, and their impact on changing maladaptive attitudes in the long term is still not well researched. Prisons in the UK and US have adopted CBT programs directed at targeting deficits in problem solving, social skills, and insight, but there is no conclusive evidence that these intervention programmes are effective in reducing recidivism.

Improvements in clinical practice with young people who offend are also urgently needed, as the effectiveness of prison sentences for young people is questioned, e.g. in the UK, around 75 per cent re-offend within two years of being released from prison (Howitt, 2018). Young Offender Institutions (YOIs) may inadvertently increase antisocial attitudes due to the pressures of peer influence, thus not necessarily always providing a prosocial experience. Therefore, imprisoning young people who have committed a criminal act, which is quite costly, may be counterproductive and not an efficient use of public funding. Welsh et al. (2008) estimated the economic costs of self-reported crimes of a cohort of young males in the 7–17 year age range in Pittsburgh, USA, as ranging from $90–110 million dollars for 500 boys. An earlier study (Yoshikawa, 1995) analysed the results of 40 pre-existing evaluation studies into social intervention projects, and argued that, in order to be more effective, interventions need to be delivered to children before they grow into 'delinquent' adolescents. They found that combined family support and early education projects showed the most promising outcomes, with improvements in cognitive abilities of children, parenting abilities, and a reduction of antisocial behaviour in the long term. For young people with the most persistent offending behaviour, projects that offered: one-to-one mentoring in order to help young people reintegrate into the school; group work addressing life problem-solving, anger management, victim awareness, interpersonal skills, substance misuse, appropriate sexual behaviour and health; and self-esteem and social skills building activities such as music, art and drama, showed the most promising outcomes, with a reduction in numbers of police charges and risk behaviours. Wilson and Hoge's (2013) more recent meta-analysis highlighted that diversion programmes were more likely to decrease re-offending rates in young people. This research seems to argue for a reduction in the use of custodial sentences in YOIs for young people, and for the use of social psychologically-informed interventions, provided by the youth justice service, perhaps as a way forward to tackle youth crime and support young people in leading safer lives. This is an approach adopted in some European countries, but not in the UK.

The Role of Cognitive Dissonance in Promoting Change

Festinger (1957) moulded the original theory of cognitive dissonance. He suggested that people who simultaneously hold inconsistent cognitions are in '*dissonance*'. This is an unpleasant, drive-like state which motivates people to change (i.e. to alter what they think to reduce the experience of dissonance). The reduction occurs by changing the cognition least resistant to change, or by adding cognitions which minimise the perceived magnitude of the discrepancy. It is comparable to other drives like thirst, hunger, sexual needs, which lead to behaviours (drinking, eating, having sex) which reduce the original drive (satisfaction). However, it is different in that dissonance is provoked solely through cognitive processes. Festinger gives the example of earthquake survivors, who tend to expect another natural disaster despite no evidence to suggest this. The survivors are still afraid, despite no longer having a reason to feel scared for their safety. Therefore they expect another earthquake for example (expectation which raises their anxieties) to produce consistency between their emotions and their thoughts.

Various conditions of dissonance arousal have been proposed in the literature. These include: (1) choice; (2) commitment; (3) responsibility; and (4) unwanted consequences. These modifiers of dissonance will be reviewed next. Linder et al. (1967) found that if people perceived that they were free to accept or decline the invitation to make a counter-attitudinal statement (i.e. they had a choice), then dissonance was aroused. Carlsmith et al. (1966), as well as Davis and Jones (1960), showed that dissonance was greater if people felt committed to their counter-attitudinal behaviour. A consistent finding in dissonance research is that it occurs when participants feel responsible for the inconsistency (Wicklund & Brehm, 1976) and when people feel responsible for the consequences of an action (please see Goethals & Cooper, 1975; Goethals et al., 1979).

The *New Look* model of dissonance (Cooper & Fazio, 1984) further developed dissonance theory by trying to make sense of all its modifiers. According to the New Look model, dissonance begins with a behaviour. In order for this behaviour to lead to a cognitive or attitude change, a set of processes that can be divided in two stages must unfold: *dissonance arousal* and *dissonance motivation*. Dissonance arousal occurs when people take responsibility for bringing about an aversive event. But there are several decision points that need to be crossed in order for a behaviour to bring about dissonance arousal. First, a behaviour has to be perceived to have an unwanted consequence for the person engaging in the behaviour, otherwise dissonance will not follow. The second decision point is the acceptance of *personal responsibility* for the consequences of the behaviour. In the New Look model, responsibility is defined as a combination of two factors: freely choosing the behaviour in question, and being able to foresee the consequences of that behaviour. In other words, accepting responsibility leads to dissonance; whereas denial of responsibility allows people to avoid the unpleasant state of dissonance. In this model (Cooper and Fazio, 1984), decision freedom and foreseeability are crucial because they are necessary for the acceptance of responsibility and, consequently, for dissonance arousal.

Stone & Cooper (2001) extended the New Look model's emphasis on aversive consequences by arguing that all behaviours have consequences.

What causes dissonance is the perception that the consequence is aversive for the self. This process requires a comparison to a standard of judgement. In their *Self-standards* model of dissonance, Stone and Cooper reasoned that there are two major categories of standards that a person can use to assess the meaning of the consequences of his/her behaviour: *normative* and *personal*. There are some outcomes in the world that most people would agree are of a particular valence (e.g. most people would agree that contributing to a charity is a positive event). Similarly, most people would agree that running into someone in the street and knocking him/her down is a negative event. When a standard of judgement is based on a perception of what most people perceive to be foolish, immoral, or otherwise negative, people are using a *normative standard of judgement*. The other broad category of standards of judgement are those that are based on the unique characteristics of the individual. These are *personal standards of judgement*. They refer solely to the judgements people make when they consider only their personal values or desires. The standard which a person uses is a function of the standard which is accessible at the time of their behaviour. In other words, if the situation makes normative standards accessible, then people will use their concept of what most people would find desirable as the way to assess the consequences of their behaviour. Conversely, if people are induced to have their personal standards accessible, then they will use their self-expectations as the standard of judgement to determine whether or not an outcome is aversive (Stone & Cooper, 2001).

The predictions of the self-standard model have been supported in a number of studies (Weaver & Cooper, 2002; Stone & Cooper, 2003). When people compare their behaviour to normative standards of judgement, then they assess consequences to be aversive in a manner similar to most people in their culture. Thus, dissonance is not moderated by their sense of self. By contrast, when idiosyncratic dissonance is aroused by comparison to personal standards of judgement, then what is considered aversive varies by self-esteem. People with a high sense of self-esteem expect to make good and rational choices. They are upset when their choices lead to a consequence that is negative. When people with chronically low self-esteem make choices, they expect those choices to have negative results, and are not upset by what other people would consider negative consequences (Weaver & Cooper, 2002; Stone & Cooper, 2003). These findings are relevant to the treatment of forensic populations. It could be argued that the current system as a whole is overwhelmingly punitive and shaming, and therefore ineffective in developing people's self-esteem, reinforcing more adaptive attitudes/behaviours, and significantly reducing re-offending rates. If forensic settings were to shift their focus from punishment to rehabilitation by focusing on decreasing stigma and increasing self-esteem through engagement in vocational, education, and occupational activities, this could in turn increase dissonance and, potentially, lead to more significant attitudes and behavioural changes. Some literature has highlighted the potentially protective effect of self-esteem on recidivism. For example, Hubbard (2008) has found that, in a sample of white male and female people with an offending history, as self-esteem levels increased, the likelihood of arrest decreased.

New avenues of dissonance research explore if dissonance within an individual can be aroused vicariously; that is, by observing someone else behaving in a counter-attitudinal fashion. This drive has stemmed from (1) the important work in social identity theory (Tajfel, 1970) and in social categorisation theory (Turner & Hogg, 1987), that suggests that people in groups forge a common identity; and from (2) recent theorising which has made it clear that the self is both personal and social (Leary & Tangney, 2003) (see Chapter 3). It is about one's own personal characteristics and, simultaneously, about one's interconnectedness with others and with a social group (Brewer & Gardner, 1996). The theory of vicarious dissonance (Cooper & Hogg, 2007) hypothesises that one group member's counter-attitudinal advocacy would cause other group members to experience dissonance vicariously, and result in attitude change by the other members of the social group. Norton et al. (2003) and Monin et al. (2004) provided evidence for vicarious arousal of dissonance, but this effect only occurred when the participants strongly identified with their group. In the absence of a strong affinity with one's group, observing an in-group member making a counter-attitudinal statement did not affect participants' attitudes.

Box 5.2 Top tips for changing attitudes and behaviour using learning from dissonance research

Maximise dissonance by:

- Seeking the client's consent to engage in interventions.
- Emphasise the client's responsibility for engagement.
- Support the client in making a commitment to change.
- Support the client with thinking about the positive and aversive consequences and their impact on the client's image.
- Expose the client to role-models (recovery champions) who are engaging in the dissonance act, in order to create vicarious dissonance.
- Raising self-esteem and self-efficacy.

The Relationship between Cognitive Dissonance and Shame

There is a growing focus in forensic settings on the use of restorative justice (RJ) approaches, which may be helpful to increase cognitive dissonance in offending populations. RJ strives to repair harm done to the victims of an offence. The main stakeholders (e.g. victim(s), person who offended) take an active role in delineating the harm caused, and developing plans to repair this. This can occur via face-to-face meetings, or by less direct means (e.g.

writing a letter). The reported powerful transformative impact of RJ approaches has been linked to Braithwaite's (1989) theory of reintegrative shaming. This distinguishes between what has been called 'stigmatic shaming' and 'reintegrative shaming'. Braithwaite describes reintegrative shaming as expressing dissatisfaction with a person in a respectful manner, which does not label them as evil, and ending in forgiveness. The theory predicts that the practice of reintegrative shaming will result in less offending. Conversely, stigmatising shaming is not respectful of the person, is not terminated by forgiveness, and labels the person as evil. Therefore the theory of reintegrative shaming argues that stigmatising shaming leads to greater levels of offending (Braithwaite 1989; Makkai & Braithwaite 1994). RJ has been embraced as a crime reduction strategy in many countries around the world (Gavrielides, 2007; Hughes & Mossman, 2001; Sullivan & Tifft, 2006). There is growing positive evidence to support its positive effects, including increased participant satisfaction, reduced re-offending rates, increased restitution compliance for forensic populations, and reduced post-traumatic symptoms for victims. In addition, there is a growing evidence base indicating that RJ may be effective for both victims and forensic populations of crimes involving serious or severe violence (Hayes, 2005; Strang, 2002; Sullivan & Tifft, 2006; Umbreit & Vos, 2000; Umbreit et al., 2003). However, research examining the use of RJ with adults who have committed violent crimes, has been criticised for having small sample sizes, and is in need of larger empirical studies (Cook et al., 2015).

Finally, cognitive dissonance and change are maximised when motivational enhancement techniques are used during therapeutic interventions (Rollnick et al., 2008). Motivational interviewing is a directive, client-centred counselling approach for initiating behaviour change by helping clients to resolve their ambivalence (cognitive dissonance). Rather than persuading the client directly, the therapist systematically elicits from the client, and reinforces reasons for concerns and for change, while maintaining a warm and supportive atmosphere for exploration of ambivalent feelings. The therapist actively avoids a confrontational approach in which they would assert the need for change while the client denies it. Responsibility for change is *owned* by the client rather than imposed by the therapist. Underlying this process is the goal of collaboratively developing a motivational discrepancy between present behaviour and desired goals. To facilitate change, it is important for therapists to understand the construction of the world from the client's perspective (i.e. identify where the client is at and meet him/her there, and then assist the client in determining the best course of action to change by helping them to set behavioural goals that are planned, observable, measurable, and relevant; Wong et al., 2007). Figure 5.1 below is a pictorial summary of the worst-case scenario, depicting the stages of entry to, and exit from, the criminal justice system (CJS), and what each stage represents in terms of potential to shame, and the subsequent reduction of dissonance opportunities. By contrast, Figure 5.2 lays out the changes that could be made, after a thorough assessment and formulation of one's offending

106 *Lara Arsuffi*

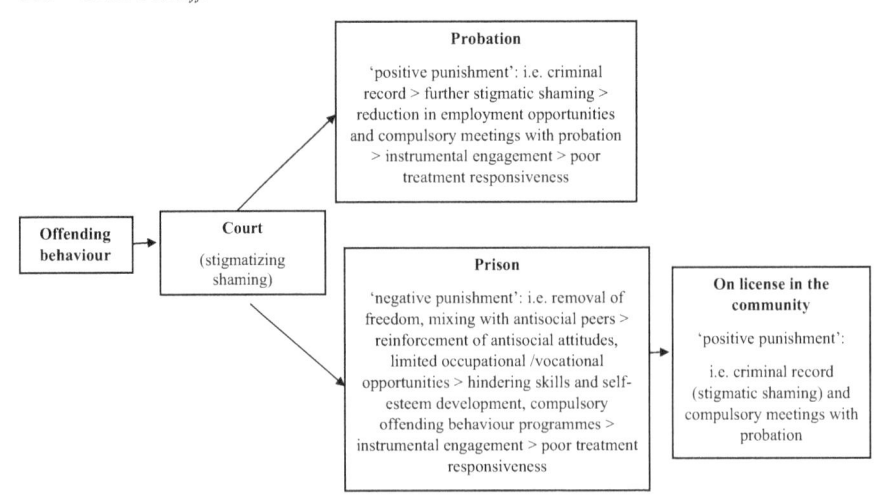

Figure 5.1 Stages of entry to and exit from the CJS with corresponding potential to shame and reduce dissonance

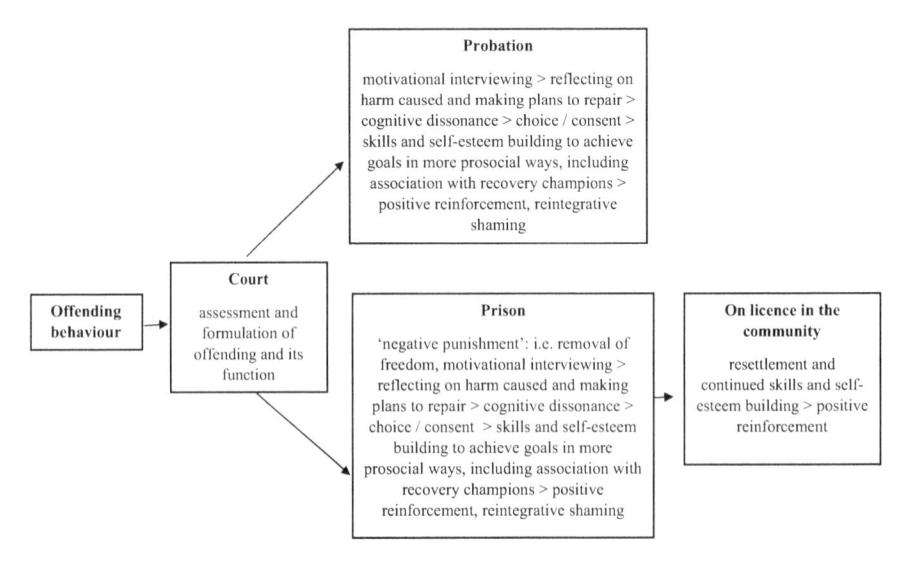

Figure 5.2 Opportunities to create dissonance and promote reintegrative shaming in the CJS

behaviour, to create opportunities to create dissonance, build self-esteem and promote reintegrative shame, as opposed to stigmatising shame, whether a person is disposed by a custodial sentence or a community order, under the supervision of probation.

So far, this chapter has argued that attitudes play an important role in people's evaluations of the social world, and it has reviewed social learning principles and how these explain the development of offending behaviour. It has also used learnings from social psychological theories to propose opportunities for improving forensic practice in order to maximise treatment effectiveness, and, ultimately, reduce recidivism rates. To conclude, this chapter will consider the role of 'schemas' and how practitioners consider the role of schemas when working with clients who have been diagnosed with personality difficulties.

What Are Schemas and How Do they Apply to Personality Theory?

Learnings from social psychology have highlighted how attitudes play a key role in the occurrence of anti-social behaviours. Building on from these insights, other literature proposes that people's attitudes are influenced by higher-order unconscious mental structures, which also need to be assessed in order to understand the predisposing factors of one's offending behaviour. For example, Young et al. (2003) argue that people's explicit and implicit attitudes about themselves, other people, and the world in general are influenced by their core beliefs and schemas. The concept of schemas has a long history in psychology. The term, which can be traced back to Piaget (1952), has been used to explain mental templates for perception, encoding, storage, and retrieval of information that integrate and attach meaning to events (Beck et al., 2004). Piaget (1952) described *"schema"* as the way that the structure of the internal world has built up through process of assimilation of ideas, and then modified as needed into consistent cognitions that form the basic cognitive system of the person. These structures help people to interpret every stimulus and maintain cognitive consistency. The idea is that experiences are stored in autobiographical memory by way of *"schemas"* formed in the early years of life (Conway & Pleydell-Pearce, 2000). Each day, people encounter a large number of stimuli from the environment. To avoid cognitive overload, humans use these organisational frameworks to be able to process the vast amount of incoming information.

In other words, these schemas are thought to form the lenses though which people look at the world. Once formed, humans have the tendency to retain their schemas, which is logical given that people strive for consistency in life. So, once people have schemas, they are likely to process information in such a way that fits with their schemas, which can eventually cause these schemas to overgeneralise (Lobbestael & Arntz, 2012). Schemas thus consist of "sensory perceptions, experienced emotions and actions, and the meaning given to them, such that early childhood experiences are memorized non-verbally" (Michiel Van & Van Vreeswijk, 2012, p. 28). Beck introduced a hierarchical model of cognition. Guiding common surface cognitions, he emphasised a middle level of conditional cognitive ideas named "core beliefs". At a deeper level of cognitive constructs, unconditional and unconscious schemas organise thought and behaviour (Kriston et al., 2012).

In recent years, cognitive behavioural theorists (e.g. Linehan, 1993; Young et al., 2003; Beck et al., 2004) have expanded their theories in order to understand and treat people with personality difficulties. Although there are differences between these theories, they all argue that personality difficulties are characterised by dysfunctional core beliefs or schemas, and that these dysfunctional core beliefs/schemas are highly inflexible, over-generalised, and lie at the core of one's self-concept. According to cognitive-behavioural theories, repeated childhood traumas and chronically maladaptive family environments may be the antecedents to the development of maladaptive schemas/dysfunctional core beliefs, which once served a protective function for the child. However, as the person grows into an adult, these maladaptive ways of perceiving the world are no longer useful, and cause the person to engage in risky behaviours, and ultimately lead to the development of personality difficulties. Personality difficulties are highly prevalent in forensic populations, and are associated with increased risk of violence and recidivism (Blackburn et al., 2003; Leistico et al., 2008). Young et al. (2003) identified a specific subset of schemas at the core of personality difficulties. He labelled these *"Early Maladaptive Schemas"* (EMS). Young proposes that schemas are different from core beliefs because they include more (unconscious) emotional memories, which are processed by different systems in the brain. Once activated, these memories can produce strong bodily sensations out of conscious awareness (Young et al., 2003). Young groups the eighteen early maladaptive schemas into five categories and names these *"schema domains"*. These are: (1) disconnection and rejection; (2) impaired autonomy; (3) impaired limits; (4) other-directedness; and (5) over-vigilance. The schema domains were defined based on unmet core emotional needs in childhood.

For example, consider a boy who grew up in a highly volatile family environment where he witnessed domestic abuse towards his mother by a succession of violent partners, alongside being emotionally neglected and abused (i.e. put down, criticised) and frequently physically assaulted. It is formulated that, based on these early life experiences, this child is likely to have formed the maladaptive schemas that the world is a dangerous and hostile place, fostering a self-protection-based approach to life, where violence is an acceptable strategy to protect himself from perceived threats and solve interpersonal problems. It is also hypothesised that this child may have become desensitised to violence, and learned maladaptive emotional coping from his role models because he was not exposed to positive problem solving and emotional management strategies, which, in turn, limited his chances of developing these coping skills himself. These experiences would have contributed to him developing implicit and explicit attitudes supporting the use of violence as a means of protecting himself from perceived risk of harm. Alongside being left traumatised by suffering physical abuse, his emotional needs were not met within the family environment. The constant criticism he was exposed to might have hindered the healthy development of his self-esteem, which may have led to the development of a sense of defectiveness (i.e. that he was not good enough to achieve personal goals, and not worthy of happy outcomes, consequently fostering a pessimistic outlook about his future). When this child enters school, he is likely to experience problems in his learning, and in relating to his peers and teachers. Without an adequate assessment of his difficulties and the provision of services to repair his views of himself and others, this child is likely to enter adolescence with many maladaptive thinking processes and skill deficits. This may, in turn, impact on his ability to become a purposeful member of society, and result in him engaging in risky behaviour to achieve his goals as this is the only way he knows how based on his childhood experiences. Consequently, he is left with a higher chance of becoming involved with the CJS compared to a child with a more positive start in life. As the adolescent grows into a young adult, his emotional, cognitive, and behavioural responses become ingrained and generalised across multiple settings (e.g. occupational, social) and may cause him (and others) significant distress, thus warranting a diagnosis of personality difficulties.

Conclusion

In summary, this chapter highlighted that implicit and explicit attitudes play an important role in our evaluations of interpersonal relationships. When distorted and maladaptive, attitudes can predispose a person to act aggressively. Therefore, these pro-criminal attitudes need to be assessed in order to understand the predisposing factors of the client's offending behaviour, and these unhelpful attitudes need to be targeted during treatment interventions. While explicit attitudes are more regularly assessed in forensic practice, assessing implicit attitudes is not yet a common procedure. Examples of implicit tests which could be used alongside explicit measures, include the Implicit Association Test (Greenwald et al., 1998), the Word Completion Task (Anderson, 1999), and the Puzzle Test (Ireland & Adams, 2015). This chapter then emphasised that people's attitudes are influenced by higher order unconscious mental structures (e.g. core beliefs and schemas), which also need to be assessed in order to understand the predisposing factors of one's offending behaviour. People who are exposed to higher rates of abuse and neglect in their developmental years, may end up offending due to these underlying beliefs being supportive of the use of violence and anti-social behaviour to protect themselves and/or achieve personal goals. Therefore, when working with forensic populations, it is important to collect a thorough account of their psychosocial history, as this information will be useful to understand what beliefs they may have of themselves, other people, and the world in general. Additionally, it may help mental health professionals to understand historical processes which have reinforced and maintained the clients' offending behaviour. These risk factors will need to be addressed during treatment provision in order to reduce the likelihood of recidivism. The final part of this chapter summarised recommendations for maximising treatment outcomes when providing interventions for forensic clients, including maximising the impact of cognitive dissonance in therapeutic programmes, recruiting recovery champions to deliver offending behaviour interventions, incorporating restorative justice ideas and exercises, building clients' self-esteem, and using motivational interviewing to roll with their resistance.

Note

1 Self-monitoring is a personality trait which relates to the ability to change our behaviour in interpersonal interaction. People low on self-monitoring use inner values when deciding how to act, while people high on self-monitoring have a tendency to change their behaviour to fit in their social environment (Snyder, 1974).

References

Akers, R. L. (1998). *Social learning and social structure: A general theory of crime and deviance*. Northeastern University Press.

Alshomrani, A. T. (2020). Addiction therapeutic communities: Residents' retention, early dropout, and their correlates over fourteen years. Retrieved from https://assets.researchsquare.com/files/rs-120410/v1/0193024e-df75-4ec7-a75e-9fccbe792c96.pdf?c=1631863198.

Anderson, C. A. (1999). Word completion task. Retrieved from https://scholar.google.com/scholar?hl=en&as_sdt=0%2C5&q=anderson+1999+word+completion&btnG=.

Andrews, D. A, & Bonta, J. (2010). *The psychology of criminal conduct* (5th edition). Matthew Bender & Co.

Bandura, A. (1969). Social-learning theory of identificatory processes. *Handbook of socialisation theory and research* (pp. 213–262). Rand McNally.

Bandura, A. (1979). *Social learning theory*. Prentice Hall.

Banse, R., Koppehele-Gossel, J., Kistemaker, L. M., Werner, V. A., & Schmidt, A. F. (2013). Pro-criminal attitudes, intervention and recidivism. *Aggression and Violent Behaviour*, 18, 673–685. doi:10.1016/j.avb.2013.07.024.

Bargh, J. A., & Pietromarco, P. (1982). Automatic information processing and social perception: The influence of trait information presented outside of conscious awareness on impression formation. *Journal of Personality and Social Psychology*, 43, 437–449.

Bargh, J. A., Raymond, P., Pryor, J., & Strack, F. (1995). Attractiveness of the underling: An automatic power-sex association and its consequences for sexual harassment and aggression. *Journal of Personality and Social Psychology*, 68, 768–781.

Beck, A. T., Freeman, A., & Davis D. D. (2004). *Cognitive therapy of personality disorder*. Guilford Press.

Blackburn, R., Logan, C., Donnelly, J., & Renwick, S. (2003). Personality disorders, psychopathy and other mental disorders: Co-morbidity among patients at English and Scottish high security hospitals. *Journal of Forensic Psychiatry and Psychology*, 14 (1), 111–137.

Braithwaite, J. (1989). *Crime, shame and reintegration*. Cambridge University Press.

Brewer, M. B., & Gardner, W. (1996). Who is the 'we'? Levels of collective identity and self-representation. *Journal of Personality and Social Psychology*, 71, 83–93.

Burgess, R. L., & Akers, R. L. (1966). A differential association-reinforcement theory of criminal behaviour. *Social Problems*, 14, 128–147.

Carlsmith, J. M., Collins, B., & Helmreich, R. L. (1966). Studies in forced compliance: 1. The effect of pressure for compliance on attitude change produced by face-to-face role playing and anonymous essay-writing. *Journal of Personality and Social Psychology*, 4, 1–13.

Carver, C., Ganellen, R., Froming, W., & Chambers, W. (1983). Modelling: An analysis in terms of category accessibility. *Journal of Experimental and Social Psychology*, 19, 403–421.

Conway, M. A., & Pleydell-Pearce, C. W. (2000). The construction of autobiographical memories in the self-memory system. *Psychological Review*, 107(2), 261–288. doi:10.1037/0033-295X.107.2.261.

Cook, A., Drennan, G., & Callanan, M. M. (2015). A qualitative exploration of the experience of restorative approaches in a forensic mental health setting. *The Journal of Forensic Psychiatry and Psychology*, 26(4), 510–531. doi:10.1080/14789949.2015.1034753.

Cooper, J., & Fazio, R. H. (1984). A new look at dissonance theory. *Advances in Experimental Social Psychology*, 17, 229–266.

Cooper, J., & Hogg, M. A. (2007). Feeling the anguish of others: A theory of vicarious dissonance. *Advances in Experimental Social Psychology*, 39, 359–403.

Davis, K. E., & Jones, E. E. (1960). Change in interpersonal perception as a means of reducing cognitive dissonance. *Journal of Abnormal and Social Psychology*, 61, 402–410.

DeBono, K. G. (1987). Investigating the social-adjustive and value-expressive functions of attitudes: Implications for persuasion processes. *Journal of Personality and Social Psychology*, 52(2), 279–287. doi:10.1037/0022-3514.52.2.279.

DeBono, K. G., & Edmonds, A. E. (1989). Cognitive dissonance and self-monitoring: A matter of context?. *Motivation and Emotion* 13(4), 259–270. doi:10.1007/BF00995538.

Dodge, K. A. (1980). Social cognition and children's aggressive behaviour. *Child Development*, 51, 162–170.

Dodge, K. A., & Crick, N. R. (1990). Social information-processing bases of aggressive behaviour in children. *Personality and Social Psychology Bulletin*, 16, 8–22.

Dodge, K. A., & Tomlin, A. (1987). Utilisation of self-schemas as a mechanism of attributional bias in aggressive children. *Social Cognition*, 5, 280–300.

Eckhardt, C. I., Samper, R., Suhr, L., & Holtzworth-Munroe, A. (2012). Implicit attitudes toward violence among male perpetrators of intimate partner violence: A preliminary investigation. *Journal of Interpersonal Violence*, 27(3), 471–491. doi:10.1177/0886260511421677.

Eliot, M., & Cornell, D. G. (2009). Bullying in middle school as a function of insecure attachment and aggressive attitudes. *School Psychology International*, 30(2), 201–214. doi:10.1177/0143034309104148.

Fazio, R. H., & Olson, M. A. (2003). Implicit measures in social cognition research: Their meaning and use. *Annual Review of Psychology*, 54, 297–327. doi:10.1146/annurev.psych.54.101601.145225.

Festinger, L. (1957). *A theory of cognitive dissonance*. Row, Peterson.

Fincham, F. D., Cui, M., Braithwaite, S., & Pasley, K. (2008). Attitudes towards intimate partner violence in dating relationships. *Psychological Assessment*, 20(3), 260–269. doi:10.1037/1040-3590.20.3.260.

Gavrielides, T. (2007). *Restorative justice theory and practice: Addressing the discrepancy*. Hakapaino Oy.

Gawronksi, B., Ehrenberg, K., Banse, R., Zukova, J., & Klauer, K. C. (2003). It's in the mind of the beholder: The impact of stereotypic associations on category-based and individuating impression formation. *Journal of Experimental Social Psychology*, 39(1), 16–30. doi:10.1016/S0022-1031(02)00517-6.

Goethals, G., & Cooper, J. (1975). When dissonance is reduced. The timing of self-justificatory attitude change. *Journal of Personality and Social Psychology*, 32, 361–367.

Goethals, G., Cooper, J., & Naficy, A. (1979). Role of foreseen, foreseeable and unforeseeable behavioural consequences in the arousal of cognitive dissonance. *Journal of Personality and Social Psychology*, 37, 1179–1185.

Greenwald, A. G., & Banaji, M. R. (1995). Implicit social cognition: Attitudes, self-esteem and stereotypes. *Psychological Review*, 102, 4–27.

Greenwald, A. G., McGhee, D. E., & Schwartz, J. L. K. (1998). *Implicit Association Test (IAT)* [database record]. APA PsycTests. doi:10.1037/t03782-000.

Hayes, H. (2005). Assessing reoffending in restorative justice conferences. *Australian and New Zealand Journal of Criminology*, 38, 77–101.

Helmus, L., Hanson, K. R., Babchishin, K. M., & Mann, R. E. (2013). Attitudes supportive of sexual offending predict recidivism: A meta-analysis. *Trauma, Violence & Abuse*, 14, 34–53. doi:10.1177/1524838012462244.

Higgins, E. T. (1996). Knowledge activation: Accessibility, applicability and salience. In E. T. Higgins & A. W. Kruglanski (eds), *Social psychology: Handbook of basic principles* (pp. 133–168). Guilford.

Higgins, E. T., Rholes, W. S., & Jones, C. R. (1977). Category accessibility and impression formation. *Journal of Experimental Social Psychology*, 13, 141–154.

HMPPS & NHS England. (2020). *Working with people in the criminal justice system showing personality difficulties* (3rd edition). HMPPS & NHS England. Retrieved from https://assets.publishing.service.gov.uk/government/uploads/system/uploads/attachment_data/file/1035881/6.5151_HMPPS_Working_with_Offenders_with_Personality _Disorder_accessible_version_.pdf.

Horgan, H., Charteris, C., & Ambrose, D. (2019). The Violence Reduction Programme: An exploration of posttreatment risk reduction in a specialist medium-secure unit. *Criminal Behaviour and Mental Health*, 29(5–6), 286–295. doi:10.1002/cbm.2123.

Howitt, D. (2018). Juvenile offenders and beyond. In D. Howitt, *Introduction to forensic and criminal psychology* (6th edition). Pearson.

Hubbard, D. J. (2008). Should we be targeting self-esteem in treatment for offenders: do gender and race matter in whether self-esteem matters? *Journal of Offender Rehabilitation*, 44(1), 39–57. doi:10.1300/J076v44n01_03.

Hughes, P., & Mossman, M. J. (2001). *Rethinking access to criminal justice in Canada: A critical review of needs and responses.* Research and Statistics Division, Department of Justice Canada.

Ireland, J. L., & Adams, C. (2015). Implicit cognitive aggression among young male prisoners: Association with dispositional and current aggression. *International Journal of Law and Psychiatry*, 41, 89–94. doi:10.1016/j.ijlp.2015.03.012.

Ireland, T. O., & Smith, C. A. (2009). Living in partner-violent families: Developmental links to antisocial behaviour and relationship violence. *Journal of Youth and Adolescence*, 38, 323–339. doi:10.1007/s10964-008-9347-y.

Keulen-de Vos, M., Bernstein, D. P., & Arntz, A. (2014). Schema therapy for aggressive offenders with personality disorders. In R. C. Tafrate & D. Mitchell (eds), *Forensic CBT: A handbook for clinical practice* (pp. 66–83). John Wiley & Sons.

Khoury-Kassabri, M. (2012). The relationship between teacher self-efficacy and violence towards students as mediated by teacher's attitude. *Social Work Research*, 36(2), 127–139. doi:10.1093/swr/svs004.

Kriston, L., Schafer, J., von Wolff, A., Harter, M., & Holzel, L. P. (2012). The latent factor structure of Young's early maladaptive schemas: Are schemas organised into domains? *Journal of Clinical Psychology*, 68(6), 684–698.

Leary, M. R., & Tangney, J. P. (2003). *Handbook of self and identity*. Guilford Press.

Leistico, A. R., Salekin, R. T., DeCoster, J., & Rogers, R. (2008). A large-scale meta-analysis relating the Hare measure of psychopathy to antisocial conduct. *Law and Human Behaviour*, 32, 28–45.

Linder, D. E., Cooper, J., & Jones, E. E. (1967). Decision freedom as a determinant of the role of incentive magnitude in attitude change. *Journal of Personality and Social Psychology*, 6, 245–254.

Linehan, M. M. (1993). *Cognitive-behavioural treatment of borderline personality disorder*. Guilford Press.

Lobbestael, J., & Arntz, A. (2012). Cognitive contributions to personality disorders. In T.A. Widiger (ed.), *The Oxford handbook of personality disorders*, 325–344. Oxford University Press.

Mahoney, J. S., Palyo, N., Napier, G., & Giordano, J. (2009). The therapeutic milieu reconceptualised for the 21st century. *Archives of Psychiatric Nursing*, 23(6), 423–429. doi:10.1016/j.apru.2009.03.002.

Makkai, T., & Braithwaite, J. (1994). Reintegrative shaming and compliance with regulatory standards. *Criminology*, 32, 361–385.

McMurran, M., Fyffe, S., McCarthy, L., Duggan, C., & Latham, A. (2001). 'Stop & Think': Social problem-solving therapy with personality disordered offenders. *Criminal Behaviour and Mental Health*, 11, 273–285.

Michiel Van, V., & Van Vreeswijk, M. (2012). *The Wiley-Blackwell handbook of schema therapy: Theory, research and practice*. John Wiley & Sons.

Monin, B., Norton, M. I., Cooper, J., & Hogg, M. A. (2004). Reacting to an assumed situation vs conforming to an assumed reaction. The role of perceived speaker attitude in vicarious dissonance. *Group Processes and Intergroup Relations*, 7, 207–220.

Nicholson, J., & Higgins, G. E. (2017). Social structure social learning theory: Preventing crime and violence. In B. Teasdale & M. S. Bradley (eds), *Preventing crime and violence: Advances in prevention science* (pp. 11–20). Springer International Publishing.

Norton, M. I., Monin, B., Cooper, J., & Hogg, M. A. (2003). Vicarious dissonance: Attitude change from the inconsistency of others. *Journal of Personality and Social Psychology*, 85, 47–62.

Piaget, J. (1952). *The Origins of Intelligence in Children*. W. W. Norton & Co.

Rollnick, S., Miller, M., & Butler, B. (2008). *Motivational interviewing in health care*. The Guilford Press.

Ross, R. R., & Fabiano, E. A. (1985). *Time to think: A cognitive model of delinquency prevention and offender rehabilitation*. Air Training & Publications.

Rudman, L. A., Ashmore, R. D., & Gary, M. L. (2001). Unlearning automatic biases: The malleability of implicit prejudice and stereotypes. *Journal of Personality and Social Psychology*, 81(5), 856–868. doi:10.1037/0022-3514.81.5.856.

Rydell, R. J., McConnell, A. R., & Mackie, D. M. (2008). Consequences of discrepant explicit and implicit attitudes: Cognitive dissonance and increased information processing. *Journal of Experimental Social Psychology*, 44, 1526–1532. doi:10.1016/j.jesp.2008.07.006.

Schneider, W., & Fisk, A. D. (1982). Degree of consistent training: Improvements in search performance and automatic process development. *Perception & Psychophysics*, 31, 160–168. doi:10.3758/BF03206216.

Schneider, W., & Shiffrin, R. M. (1977). Controlled and automatic human information processing: I. Detection, search, and attention. *Psychological Review*, 84(1), 1–66. doi:10.1037/0033-295X.84.1.1.

Sempertegui, G. A., Karreman, A., Arntz, A., & Bekker, M. H. J. (2013). Schema therapy for borderline personality disorder: A comprehensive review of its empirical foundations, effectiveness and implementation possibilities. *Clinical Psychology Review*, 33, 426–447. doi:10.1016/j.cpr.2012.11.006.

Skinner, B. F. (1953). *Science and human behaviour*. Macmillan.

Skinner, B. F. (1979). *The shaping of a behaviourist*. Alfred A. Knopf.

Smith, C. A., Ireland, T. O., & Thornberry, T. P. (2005). Adolescent maltreatment and its impact on young adult antisocial behaviour. *Child Abuse and Neglect*, 29, 1099–1119. doi:10.1016/j.chiabu.2005.02.011.

Snyder, M. (1974). Self-monitoring of expressive behaviour. *Journal of Personality and Social Psychology*, 30, 526–537. doi:10.1016/j.jesp.2009.09.004.

Stewart, R. L., Latu, I. M., Kawakami, K., & Myers, A. C. (2010). Consider the situation: Reducing automatic stereotyping through situational attribution training. *Journal of Experimental Social Psychology*, 46(1), 221–225.

Stone, J., & Cooper, J. (2001). A self-standards model of cognitive dissonance. *Journal of Experimental Social Psychology*, 37, 228–243.

Stone, J., & Cooper, J. (2003). The effect of self-attribute relevance on how self-esteem moderates attitude change in dissonance processes. *Journal of Experimental Social Psychology*, 39, 508–515.

Stouthamer-Loeber, M., Loeber, R., Hormish, D. L., & Wei, E. (2001). Maltreatment of boys and the development of disruptive and delinquency behaviour. *Development and Psychopathology*, 13, 941–955.

Strang, H. (2002). *Repair or revenge: Victims and restorative justice*. Oxford University Press.

Sullivan, D., & Tifft, L. (2006). *Handbook of restorative justice: A global perspective*. Routledge.

Sutherland, E. H. (1947). *Principles of criminology* (4th edition). J. B. Lippincott.

Tajfel, H. (1970). Experiments in intergroup discrimination. *Scientific American*, 223, 96–102.

Todorov, A., & Bargh, J. A. (2002). Automatic sources of aggression. *Aggression and Violent Behaviour*, 7(1), 53–68.

Turner, J. C., & Hogg, M. A. (1987). *Rediscovering the social group*. Blackwell.

Umbreit, M. S., & Vos, B. (2000). Homicide survivors meet the offender prior to execution: Restorative justice through dialogue. *Homicide Studies*, 4, 63–87.

Umbreit, M. S., Vos, B., Coates, R. B., & Brown, K. (2003). Victim offender dialogue in violent cases: The Texas and Ohio experience. *VOA Connections*, 14, 12–17.

Weaver, K. D., & Cooper, J. (2002). *Self-standard accessibility and cognitive dissonance reduction*. Paper presented at the Meeting of the Society for Personality and Social Psychology, Palm Spring, CA, January.

Welsh, B. C., Loerber, R., Stevens, B. R., Stouthamer-Loeber, M., Cohen, M. A., & Farrington, D. P. (2008). Cost of juvenile crime in urban areas: A longitudinal perspective. *Youth Violence and Juvenile Justice*, 6(1), 3–27.

Wicklund, R. A., & Brehm, J. W. (1976). *Perspectives on cognitive dissonance*. Erlbaum.

Widom, C. S. (1989a). Child abuse, neglect and violent criminal behaviour. *Criminology*, 27, 251–271. doi:10.1111/j.1745-9125.1989.tb01032.x.

Widom, C. S. (1989b). The cycle of violence. *Science*, 244, 160–166. doi:10.1126/science.2704995.

Widom, C. S., & Maxfield, M. G. (2001). *An update on the cycle of violence*. US Department of Justice, Office of Justice Programs, National Institute of Justice.

Wilson, T. D., & Brekke, N. (1994). Mental contamination and mental correction: Unwanted influences on judgements and evaluations. *Psychological Bulletin*, 116, 117–142.

Wilson, T. D., & Hoge, R. D. (2013). The effect of youth diversion programs on recidivism: A meta-analytic review. *Criminal Justice and Behaviour*, 40(5), 497–518.

Wong, S. C. P., & Gordon, A. (2013). The Violence Reduction Programme: A treatment programme for violence-prone forensic clients. *Psychology, Crime & Law*, 19(5–6), 461–475. doi:10.1080/1068316X.2013.758981.

Wong, S., Gordon, A., & Gu. G. (2007). Assessment and treatment of violence-prone forensic clients: An integrated approach. *British Journal of Psychiatry*, 190(49), 66–74. doi:10.1192/bjp.190.5.s66.

Yoshikawa, H. (1995). Long-term effects of early childhood programs on social outcomes and delinquency. *The Future of Children*, 5(3), 51–75.

Young, J. E., & Brown, G. (2003). *Young schema questionnaire*. Professional Resource Exchange.

Young, J. E., Klosko, J. S., & Weishaar, M. E. (2003). *Schema therapy: A practitioner's guide*. Guilford Press.

Zajonc, R. B. (1968). Attitudinal effects of mere exposure. *Journal of Personality and Social Psychology*, 9(2), 1–27.

Zelli, A., Cervone, D., & Hussman, L. R. (1996). Behavioural experience and social influence: Individual differences in aggressive experience and spontaneous versus deliberate trait inference. *Social Cognition*, 14, 165–190.

Zelli, A., Hussman, L. R., & Cervone, D. (1995). Social inference and individual differences in aggression: Evidence for spontaneous judgements of hostility. *Aggressive Behaviour*, 21, 405–417.

6 Aggression

Matt Bruce and Veronica Rosenberger

Overview

This chapter aims to provide an overview of social psychology's contribution to our understanding of aggression and its application to forensic practice. The fabric of social psychology stretches across many areas of the criminal justice system, and its fibres are woven into many intervention programmes that flourish in prisons, community probation services, court diversion programmes, forensic psychiatric hospitals, and youth offending teams. The focus of this chapter will be the examination of social psychological principles within forensic mental health settings. Case examples and practice exercises will also be used to illustrate the application of these principles. First, the contribution of social psychology to the understanding of human aggression will be outlined, together with a working definition of the term. Next, various typologies and theories of aggression will be discussed and applied. The chapter will then focus on approaches to assessment, prediction, and management of aggressive behaviours within forensic settings. Finally, an adapted version of a dynamic risk assessment, underpinned by social psychological principles, will be introduced for use in forensic inpatient settings. This adapted version attempts to weave together the essential elements of assessment, formulation, intervention, and reflective practice in mitigating aggressive acts among inpatients and staff.

Social Psychology and Aggression

Social psychology refers to the examination of human behaviours based on interactions that take place within a social context. It presupposes that all behaviour is the result of social cues, learning environments, and interactions with others. Accordingly, the common thread that ties all social psychological theories of aggression together is the emphasis placed on the social context in which aggressive behaviours occur. Furthermore, aggression and violence can be understood as a tool for managing interpersonal relationships and interactions- to punish, manipulate, and control the behaviours of others for the purposes of protection, self-preservation, gratification, or goal acquisition.

DOI: 10.4324/9781315560243-6

The collection of social psychological approaches to understanding aggression are extensive, and include frustration-aggression; situational determinants of the strength of the instigation to aggression; factors affecting inhibition of aggression; determinants of the nature and target of aggressive response to frustration; displacement of aggression in intergroup conflicts and scapegoating; aggression aroused by opposing belief systems and competitive situations; the effects of violence in television and movies; characteristics and development of aggressive personalities; as well as antisocial acts of aggression such as crime, suicide, and homicide (Berkowitz, 1962).

Definitions and Typologies of Aggression

Definitions of Aggression

Despite the ubiquitous and harmful effects of aggression and violence on the individual, group, and wider society, these two terms remain poorly defined and misunderstood. Some authors appear to use the terms interchangeably:

> violence and aggression refer to a range of behaviours or actions that can result in harm, hurt or injury to another person, regardless of whether the violence or aggression is behaviourally or verbally expressed, physical harm is sustained or the intention is unclear.
> (National Institute for Health and Care Excellence, 2015, p. 19)

Yet others assert a clear distinction: "aggression refers to the intention to hurt or gain advantage over people without necessarily involving physical injury" and "violence involves the use of strong physical force against another person, sometimes impelled by aggressive motivation" (Howells & Hollin, 1989, p. 4). Other theorists see aggression and violence on a continuum of 'harm', with violence located at the more extreme end of the spectrum.

Many social psychologists prefer to define aggression simply and clearly as "behavior that is intended to harm another individual who does not wish to be harmed" (Baron & Richardson, 1994, p. 8). The use of the word *intention* in many definitions of aggression places the importance of perception of intent (which is arguably contingent on the social context and the observer) at centre stage. While the authors accept that the term aggression is not always associated with maladaptive or harmful intent and/or outcomes, and that prosocial and social sanctioned uses of aggression exist, for the purposes of clarity and utility, this chapter will adopt the definition provided by Baron and Richardson (1994) unless otherwise stated.

Typologies of Aggression

Over the decades, aggression has been classified using an abundance of frameworks and nomenclature. It may be classified by the target or victim (e.g. self-directed, staff, or co-patient directed), mode of expression (e.g. physical or verbal, direct or indirect), desired or actual outcome (e.g. positive, healthy, and prosocial vs negative, harmful, and antisocial), as well as the cause or motive (e.g. impulsive vs planned). In the following section, both dichotomous and multiple classification models for categorising aggressive behaviours will be outlined.

The reactive and proactive dichotomy represents perhaps the most established yet debated classification system in the aggression literature (Bushman & Anderson, 2001). Box 6.1 lists the most prevalent and accepted dichotomous terms for reactive and proactive aggressive behaviour used in forensic settings today.

Reactive typologies of aggression are typically characterised by high levels of autonomic arousal and associated with negative emotions such as anger or fear. These forms of aggression usually represent a response to perceived frustration, stress, or threat. When a threat is dangerous and imminent, this unpremeditated aggression may be considered defensive and thus part of the normal repertoire of human behaviour. Alternatively, proactive typologies of aggression represent planned or premeditated behaviour that is not typically associated with heightened emotions, frustration, or response to immediate threat. This form of aggression is sometimes referred to as predatory or instrumental. Proactive aggression is usually mobilised with clear objectives and goals to avoid or acquire something. Proactive typologies can also be socially sanctioned, such as in wartime, or to defend property, possessions, or persons, or prevent a social wrongdoing.

While the dichotomous approach to classifying aggressors has been popular, it has also attracted increasing criticism for being overly simplistic and mutually exclusive. Indeed, the more recent concept of the "mixed motive" aggressor has challenged these single motive conceptualisations (Gendreau & Archer, 2005). Furthermore, these two classifications do not adequately distinguish pleasure- and fear-induced aggression (Chichinadze et al., 2011), consider the role of the mental state of the aggressor (Nussbaum et al., 1997; Nolan et al., 2003; Urheim et al., 2014), or identify specific motives for aggressive behaviour (Daffern et al., 2007; Bruce, 2014).

Box 6.1 Aggression dichotomies

Reactive	*Proactive*
Impulsive	Premeditated
Hostile	Instrumental
Expressive	Impassive
Defensive	Offensive
Emotional	Rational
Hot-headed	Cold-blooded
Affective	Predatory

Researchers have attempted to address these criticisms and propose multifactorial classification systems. Chichinadze et al. (2011) offers a four-part typology: fear-induced reactive defensive aggression; anger- or frustration-induced offensive aggression accompanied by anger; instrumental type of offensive aggression; and aggression linked to pleasure. With respect to the mental state of the aggressor, Nussbaum et al. (1997) postulated a three-factor model of aggression: irritable, instrumental, and defensive types which has demonstrated utility within forensic inpatient settings (Urheim et al., 2014). Nolan et al. (2003) described an alternative three-factor classification for aggressive incidents on psychiatric units: impulsive; planned; and psychotic. Quanbeck et al. (2007) developed a similar triad which they divided into impulsive, organised, and psychotic. Presumably, the latter two classification systems augmented the established reactive and proactive distinction by adding a "mental illness driven" category. However, these trichotomous approaches, while developed for individuals in psychiatric care, continue to lack specificity with respect to the function and motive behind aggression. In response to these shortcomings, Daffern et al. (2007) developed the Assessment and Classification of Function (ACF) to classify ten distinct functions (or motives) for aggressive behaviour exhibited by forensic psychiatric inpatients, which included demand avoidance, reduction of tension, obtain tangibles, status enhancement, and pleasure. The authors concede that one aggressive act may serve multiple functions (that may be deliberate and accidental). This limits the utility of the ACF if many of its categories conflate with one another. Furthermore, despite victimisation, fear, and vulnerability being a common experience for individuals with severe mental illness (Walsh et al., 2003; de Mooij et al., 2015), none of the ten functions directly address fear-mediated aggression.

Consider the case example in Box 6.2, which describes the behaviour of an inpatient during a day shift at a busy forensic inpatient unit. Using this extract, first try to identify the various possible forms of aggressive behaviours described and then explore possible typologies.

Box 6.2 Nursing note regarding Maurice

Maurice has not been interacting with his peers as usual today, and instead pacing the unit corridors. When approached by a member of staff, he appeared agitated and distracted. He refused to return to the common area and stated that he was waiting for his visit. During a patient incident, he refused to stay in his room. He was also reported to have become hostile and disrespectful when told to refrain from using the patient telephone, as it was not working. Despite these instructions, he repeatedly made attempts to use it. He continually hurled insults and profanities at staff. His visitor did not arrive during the day shift as planned. At 7.10 p.m., without warning, Maurice approached the telephone and struck the receiver against the wall, causing it to shatter. He was instructed to return to his room and continue with his programme homework.

Theories of Aggression

Foundational Theories

Historically, aggression was more or less viewed as a homogeneous category of behaviour (Kempes et al., 2005) until two competing theories grew to dominate the study of human aggression: the frustration-aggression hypothesis (Berkowitz, 1989) and the social learning theory (Bandura, 1973). This theoretical split eventually evolved into what is now considered the reactive proactive taxonomy of human aggression (Merk et al., 2005). The social psychological theories that follow each theoretical doctrine are outlined in Table 6.1. Notably, overtime, researchers have incorporated these seemly disparate theoretical principles into more integrated conceptualisations of aggression.

Integrated Theories

The Social Information Processing (SIP) theory analyses how cognitive interpretation of social interactions can impact behaviour and, potentially, inform aggression. SIP was developed in the late 1980s by two groups of researchers who elaborated the theory in distinct directions. Huesmann (1988), as described above, developed a schema- or script-based approach to explaining aggressive behaviour, while Dodge et al. (1986) considered social cognition a real-time and conceptual process. Dodge et al. (1986, 1994) argue that aggressive behaviour is an immediate response to social cues and involves an encoding process characterised by five key steps. Reactive forms of aggression have been linked with problems at the first two stages of information processing- encoding and interpretation of cues- whereas proactive types of aggression are more commonly associated with problems in the final three steps of information processing: clarification, response access, and response decision (Kempes et al., 2005). Hostile attribution bias has also become a significant pillar in SIP theory and is defined as the tendency to perceive ambiguous events as the result of hostile intentions (Warburton & Anderson, 2015). Ultimately, SIP theorises that aggressive behaviour is the result of faulty cognitive processing and that these information processing problems occur differently for different individuals (Merk et al., 2005).

The general aggression model (GAM; Anderson & Bushman, 2002) was designed to incorporate a multitude of other social psychological theories of aggression into a unified whole. It focuses on streamlining theories to increase application and develop more comprehensive interventions for individuals who are chronically aggressive (Anderson & Bushman, 2002). One of the foundations of GAM, based on the research by Berkowitz (1989, 1990) and his cognitive neoassociation theory, is the idea that brain connectivity, or knowledge structures, inform all human responses to stimuli

Table 6.1 Key features of various reactive and proactive theories of aggression

Theory	Theorist	Key Principles	Criticisms
Reactive Theories			
Frustration-Aggression Hypothesis	Dollard et al. (1939)	Aggression results from frustration, caused by an externally generated interruption in the process of goal-directed activity.	Overemphasis on frustration as mediating mechanism for aggression. Does not directly address instrumental use of aggression.
Excitation-Transfer Theory	Zillmann (1971)	Emotional arousal (and aggressive responses) to stimuli is intensified by exposure to prior stimuli not directly related to the original stimulus.	Does not explain individual differences in anger expression. Heavy focus on physiological arousal as key mediator for aggressive expressions.
Cognitive Neoassociation Theory	Berkowitz (1990)	Aversive events trigger a network of aggression-related thoughts and feelings which interact with person-situational factors leading to fight/flight responses.	Retains emphasis on frustration as key trigger of aggressive responses. Lack of emphasis on displaced and/or instrumental aggression.
Proactive Theories			
Social Learning Theory	Bandura (1973)	Aggression is a learnt via observation, imitation, and modelling of others which is subsequently reinforced by rewards (witnessed or directly experienced)	Ignores the role of emotional arousal and frustration. Minimises reactive and impulsive elements of aggression.
Script Theory of Aggression	Huesmann (1988)	Aggressive behaviour is the result of readily available mental scripts (or action templates) previously established via direct or observational learning.	Little explanation regarding how scripts are acquired. Unclear how this cognitive construct differs from schema theory and associated limitations.
Social Interactionist Theory	Felson & Tedeschi (1993)	Aggression is an instrumental behaviour mobilised in response to perceived situational or interpersonal conflict, to exert social influence, express grievances and/or maintain desired identities.	Considers aggression to be solely attributable to goal acquisition. Overemphasis on social context of aggressive expressions.

(Breuer & Elson, 2017). These knowledge structures develop from personal experiences, influence perception of stimuli, can lead to automatic behaviours, are linked to affective states and beliefs, and are, ultimately, used to guide situational interpretation and behavioural response (Anderson & Bushman, 2002). Because GAM draws on multiple theories, it can explain the widest range of aggressive behaviour, which includes both reactive and proactive categories. By focusing on how personal and situational variables impact cognitive, affective, and arousal states, GAM has been argued to produce a very individualised approach to aggression analysis (Warburton & Anderson, 2015). While GAM appears to offer a more holistic approach to our understanding of aggressive behaviour, the evidence base for this model remains heavily rooted in the laboratory and thus its generalisability to real world remains debated (Elson & Ferguson, 2014). Additionally, the GAM's focus on cognitive scripts leaves little room for understanding the motivational and characterological variables that influence aggressive behaviour.

In summary, accurate identification of these potentially distinct typologies and pathways to aggression is vital in determining appropriate and effective interventions. For example, reactive aggressors may require interventions targeting emotional management, impulse control, and social problem solving, while proactive aggressors are likely to benefit more from positive role modelling and exposure to alternative reinforcement contingencies.

Assessment of Aggression

The relationship between aggression theory and assessment is stormy. While some assessment techniques for aggression actively embrace theory, others reject it altogether. Perhaps the conceptually dense, notoriously dynamic, and multi-faceted nature of aggression has contributed to this rocky courtship. There are two critical issues that require careful consideration when identifying and selecting appropriate tools for the assessment of aggression: (i) the purpose (e.g. to define its nature, understand function, predict future likelihood); and (ii) the approach (e.g. individualised or group-based tools).

Assessment Purposes

The suite of tools designed to aid practitioners assess aggression is extensive. Over 200 structured instruments are currently available (Douglas et al., 2014), each for a specific purpose, time frame, population, and type of aggressor. Careful consideration should be given to these various areas in order to ensure the most reliable and valid assessment tool is selected for use. It is the practitioner goal to: define its nature; understand its motive or function; predict its likelihood; develop management or care plans; categorise risk levels; develop a treatment plan; or determine suitability for discharge/release. Arguably these purposes fall into two broad assessment groups: retrospective and prospective.

Retrospective assessments require the assessor to determine why an act of aggression may have occurred (i.e. offence formulation). Prospective assessments involve the assessor determining the likelihood of future aggression and detailing its probable nature, function, and timing (i.e. re-offence prediction; Bruce, 2014). Prospective assessments of aggressive behaviour have been argued to fall into three categories or generations. The first is unstructured clinical opinion that involves determining risk of future aggression according to the experience and qualifications of the professional making them. The second is the actuarial approach that uses algorithmic, norm-based reference groups to provide probabilistic estimates for likelihood of future aggressive behaviour. These assessment tools (e.g. Static-99R; Phenix et al., 2016) consider only static (unchangeable) risk factors and require no clinical judgment or theory. The third is the structured professional judgment approach, or guided clinical approach (e.g. Historical, Clinical, Risk Management-Version 3; Douglas et al., 2014) which use established theoretical, expert, and empirical knowledge. These assessments consider static, dynamic (changeable), and protective factors in risk determination, and attempt to bridge the gap between unstructured clinical opinion and actuarial methods. These latter tools also represent a hybrid of retrospective and prospective approaches to assessment.

The utility of established structured tools has received increased attention in recent years (Lantta et al., 2016; Ramesh et al., 2018) in part due to the time frame with which many are designed to cover. Many actuarial and structured professional judgment risk assessments, which have amassed a large body of empirical support and validation, have been developed to assess the likelihood of reoffending among people who have committed offences (i.e. those who have committed at least one prior aggressive act that has resulted in a criminal conviction) over the course of years. In fact, most actuarial tools are normed for predicting aggressive acts decades from the date of assessment. Accordingly, these tools have little utility when it comes to making short-term (e.g. months) or imminent (e.g. days or hours) predictions (Ogloff & Daffern, 2006). A recent systematic review by Ramesh et al. (2018) identified a number of validated and reliable actuarial tools designed to make short-term and imminent predictions of aggression and include: (i) Broset Violence Checklist (BVC; Almvik et al., 2000); (ii) Classification of Violence Risk (COVR; Monahan et al., 2005); (iii) Violence Risk Screening-10 (V-RISK-10; Bjørkly et al., 2009); and (iv) Dynamic Appraisal of Situational Risk (DASA; Ogloff & Daffern, 2003).

Assessment of aggression in forensic psychiatric settings must also take into account the dynamic and fluctuating nature of acute mental illness. Unfortunately, actuarial tools that use static and unchangeable factors alone to determine risk are unable to account for such clinical phenomena. Due to these shortcomings, Ogloff and Daffern (2006) question their utility for the purposes of risk assessment for aggression. Importantly, they highlight the distinction between risk *status* and risk *state*. While risk status (e.g. categorising individuals as high, medium, low) is useful for making longer-term predictions and assisting placement and/or discharge planning, they contend that risk state (e.g. categorising individuals based on fluctuations in their mental state) affords greater utility within inpatient settings.

Furthermore, individuals residing within inpatient settings have been argued to exhibit higher rates of reactive, compared to proactive, forms of aggression (McDermott et al., 2008; Quanbeck et al., 2007). However, it remains unclear whether individuals residing within secure forensic settings display disproportionate rates of reactive aggression. There is some evidence to suggest that, while reactive aggression has been associated with symptoms of mental illness and the clinical and risk management items on the HCR-20v3, proactive/instrumental aggression has been correlated with psychopathy (Vitacco et al., 2009; McDermott et al., 2008). Accordingly, forensic inpatients may require short-term and imminent risk assessment for predicting both reactive and proactive types of aggression.

Finally, research indicates that reactive aggression is consistently more difficult to predict than proactive aggression (McDermott & Holoyda, 2014). One explanation for this discrepancy in predictive validity is that reactive aggression relates less to static factors and more to dynamic factors that are, by definition, in a constant state of change. Another interpretation is that reactive aggression observed within inpatient settings is inherently linked to staff and environmental characteristics that represent important determinants, which are often overlooked in psychiatric inpatient settings (Johnson, 2004; Cutcliffe & Riahi, 2018).

Assessment Approaches

There are two major approaches to the assessment of aggression: 'idiographic' (e.g. generating individualised, tailored, person-centred risk profiles and offence formulations) and 'nomothetic' (e.g. establishing general tendencies, group-derived norms or categories, and re-offending probabilities). Social psychological theories have played a pivotal role in the formation of a number of significant idiographic and nomothetic approaches to assessment of aggression. Megargee's (1982) Algebra of Aggression and the Dynamic Appraisal of Situational Aggression (Ogloff & Daffern, 2003) represent two excellent contributions from the field of social psychology and provide examples from each major canon.

Megargee's (1982) Algebra of Aggression represents an idiographic pseudo-algebraic formula that places the concept of 'choice' at centre stage. He argues that at any given time, multiple responses, both aggressive and non-aggressive, are competing for expression. By means of an "internal algebra", an individual will calculate and choose the response that offers the most satisfaction at the least cost. This 'reaction potential' (or net strength) of this response will be determined by subtracting those factors deterring it, from those promoting an aggressive response (Bruce, 2014). Megargee argues that there are three personal and two situational factors that determine the reaction potential for any given aggressive act. The first personal factor is Instigation to Aggression (A) that promotes aggressive responses and is divided into 'intrinsic' and 'extrinsic instigation to aggression'. Intrinsic instigation is rooted in frustration-aggression and cognitive neoassociation theory (Berkowitz, 1993), and relates to reactive aggression. Extrinsic (instrumental) instigation is related to social learning theory (Bandura, 1973), which serves to achieve desired goals such as obtaining tangibles, self-esteem, power, social status, and relates to proactive aggression. Accordingly, this first personal factor is closely aligned to the social information processing theory of aggression. The second personal factor is 'Habit Strength' which concerns the extent to which aggressive responses have previously been rewarded or punished, thus relating to social learning theory (Bandura, 1973), aggressive script theory (Huesmann, 1998) as well as operant conditioning principles (Skinner, 1963). The final personal factor is 'Inhibitions against Aggression' which represents all reasons why an individual might refrain from acting aggressively, such as moral or practical considerations, and will vary as a function of the act, target, and circumstance (Megargee, 2009). Situational factors are divided into those that encourage the use of aggression and those that inhibit it, and include environments, settings, contexts, and situations. This social psychological framework for assessing aggression allows for structured clinical formulations of risk, however it does not offer any empirical data with respect to its predictive validity. A critique of this assessment model is that it requires rather sophisticated conceptual and clinical expertise to complete, and it remains unclear how it can be used to determine imminent risks of aggression or its applicability to individuals with severe mental illness.

A promising nomothetic approach to the assessment of aggression, which addresses some of the limitations of Megargee's approach, is the Dynamic Appraisal of Situational Aggression (DASA; Ogloff & Daffern, 2006). This instrument was specifically designed for imminent (24 hour), context specific, clinical assessment of risk within forensic psychiatric inpatient settings. This actuarial instrument was developed by combining the best short-term predictors of physical aggression from other well established and empirically validated instruments, including two items from the Broset Violence Checklist (BVC; Almvik et al., 2000) and two items from the HCR-20 (Webster et al., 1997) as well as other empirically derived items identified by Ogloff & Daffern (2006). The DASA has demonstrated superior predictive accuracy for violence in the 24 hours post assessment (Ramesh et al., 2018), and exceeded those predictions by nursing staff (Griffith, et al., 2013; Lantta et al., 2016). It comprises of seven dynamic behavioural items (see Table 6.2) coded on a two-point scale (0 = absent, 1 = present) with respect to the previous 24 hours. The DASA provides three risk categories (0 = low, 1–3 = moderate, ≥4 = high risk) based on the number of dynamic behavioural items that were present in the past 24 hours (Ramesh et al., 2018).

Table 6.2 DASA dynamic behaviour items and examples

	Behaviour Item	Example
1	Irritability	Easily annoyed or angered, unable to tolerate the presence of others.
2	Impulsivity	Hour-to-hour fluctuations in mood, thinking, and general demeanour. Displays sudden and unpredictable behaviours.
3	Unwillingness to follow directions	Refusing to be redirected, adhere to treatment, or engage in alternative behaviours.
4	Sensitivity to perceived provocation	Misinterpreting others' actions as threatening, deliberate, or harmful. Acting overly sensitive and quick to anger.
5	Easily angered when requests are denied	Refusal to accept delays in gratification of perceived needs, becoming frustrated when goals are not immediately obtained.
6	Negative attitudes	Expression of pro-criminal, disrespectful, hostile, or anti-authority attitudes towards others.
7	Verbally threatening	A verbal outburst which has a definite intent to intimidate or threaten another person.

Consider once again the nursing note in Box 6.2. Which dynamic behavioural items are relevant and present for Maurice? If tasked with scoring the DASA for this individual in care, what risk category do you think he would likely fall into for the potential for aggression in the subsequent 24 hours?

Barry-Walsh et al. (2009) found that among individuals with severe mental illness, each increase in total DASA score was associated with a 1.77 times increased likelihood that the patient would behave aggressively in the following 24 hours. Ogloff & Daffern (2006) suggest that scores of 1 or more should mobilise preventative measures, and scores of 6–7 require imminent prevention. However, the DASA, like many other nomothetic tools, lacks both theoretical foundations and subsequent management directions (Hipp et al., 2020; Kaunomäki et al., 2017). Furthermore, guidance concerning the nature or function of preventative measures remains poorly articulated. Only one study examined preventative measures following observations of high DASA scores and found that the most frequent prevention was psychopharmacological or coercive (i.e. seclusion; Kaunomäki et al., 2017). These findings most likely reflect the DASA's lack of theoretical underpinnings from which to guide psychosocial interventions.

Interventions for Aggression

We use the term 'interventions' to refer to a combination of programmes designed to prevent, treat and/or manage pathological or harmful uses of aggressive behaviour. While a substantive evidence base regarding the effectiveness of programmes designed to reduce aggressive and violent behaviour among incarcerated people exists (Gilbert & Daffern, 2010), the same cannot be said for interventions delivered in the community (Bruce, 2014) or within psychiatric settings. However, some of the most well-established and theoretically grounded programmes currently in operation to reduce aggressive and violent behaviour are rooted within social psychological principles. These interventions can be broadly divided into single mode, multimodal, and dynamic management programmes.

Single Mode Therapies

These typically address one core construct in the expression of aggressive behaviour. Perhaps the most widely accepted and utilised interventions are anger management programmes which are founded on the principles espoused by Novaco's stress inoculation-coping skills approach (Novaco, 1975, 1977). These programmes include increasing self-awareness of anger traits and states, identifying triggers, and providing a range of coping strategies, social skills training, and relaxation training. Specific examples include Controlling Anger and Learning to Manage It – Effective Relapse-Prevention Program (CALMER). This programme targets internal and external factors that trigger anger, arousal-reduction techniques, thoughts that lead to problematic behaviour, assertive communication skills, and development of a relapse-prevention plan (Winogron et al., 2001). Aggression Replacement Training (ART), originally designed for young, assaultive, quick-tempered male adolescents who demonstrate impulse control difficulties, has since been revised for adults and comprises of three key components; social skills, anger control training, and moral reasoning with evidence demonstrating its efficacy (Goldstein et al., 2004; Hatcher et al., 2008).

These programmes have received mixed support with respect to their effectiveness, notably due to their focus on reactive aggressors. Bruce (2014) outlined the following issues with single mode interventions: their narrow and insufficient understanding of emotions and cognition; a lack of specificity to tackle entrenched aggressive scripts; inadequate programme duration and intensity ('dosage'); and a limited coverage of multiple criminogenic needs that underpin violent offending. Accordingly, these deficiencies have led to the development of programmes that address multiple causes of aggressive acts, including proactive sources which reflect social learning, and entrenched normative beliefs and aggressive behavioural scripts, rather than, exclusively, impulse control deficits and anger (Polaschek & Reynolds, 2004; Gilbert & Daffern, 2010). Indeed, Novaco (2013) has conceded that "the provision of anger control treatment is an adjunctive therapy [and] best done as part of a multifaceted treatment programme" (p. 211).

Multimodal Therapies

These interventions aim to address a multitude of factors to reduce the use of aggression and violence. These include Therapeutic Communities (Jones, 2013); Violent Offenders Treatment Programme (VOTP; Braham et al., 2008); and Multi-Systemic Therapy (MST; Borduin, 1999; Henggeler et al., 2009). Therapeutic communities embrace social ecological, social learning, and humanistic principles as core agents for change (Vandevelde et al., 2004). The core concept of these residential environments is "living learning" and developing alternative, non-aggressive, methods of operating in a community. However, the effectiveness of therapeutic communities remains controversial (Malivert et al., 2012) and lacks empirical support (Jolliffe & Farrington, 2007), specifically for proactive aggressors (Jones, 2013). The VOTP targets three categories of aggressive behaviour: symptom related (psychotic); emotional reactions (reactive); personality driven (proactive); and embraces social learning principles, emotional regulation strategies, social problem solving, and cognitive therapy (Braham et al., 2008). However, as this programme was developed for forensic patients in high security psychiatric hospitals, it lacks consideration of wider systemic variables related to aggressive behaviours. Multisystemic therapy represents a systems approach that considers multiple social factors contributing to risk of aggressive behaviour, including school, work, peer-groups, family, and community factors. Subsequent interventions aim to restrict opportunities to observe or engage in aggressive behaviours and promote exposure and reward involvement in non-aggressive prosocial behaviours (Anderson & Bushman, 2002) as well as linking families to individual, school, and community support (Malivert et al., 2012).

While these multimodal interventions represent the gold standard of evidence-based treatments, they are not without their limitations. They require extensive resources, strict adherence to administration protocols, participation of those assessed as a high-risk of offending only, and delivery by highly trained professionals in group settings. Such requirements are rarely available on busy forensic psychiatric units. Instead, there is a need for day-to-day, dynamic interventions that can be delivered by a range of healthcare professionals in response to an array of fluctuating risk levels and aggressive incidents exhibited by patients.

Dynamic Management

Within this chapter, the term 'dynamic management' refers to interventions that can be employed by an array of healthcare professionals with minimal (if any) specialist psychological training, and without the need for structured group therapy. Daffern, Howells, and Ogloff (2007) have offered a systematic framework for addressing aggressive incidents on psychiatric wards. Specifically, they offer nine intervention strategies that can be employed by nursing staff and other healthcare assistants which map onto the nine identified functions of aggressive acts (detailed previously). Owing to the overrepresentation of reactive aggressive incidents observed on psychiatric units, they argue that anger management techniques, with their emphasis on reducing emotional arousal, relaxation, avoidance of high-risk situations, restructuring dysfunctional or distorted beliefs and assertion training (Renwick et al., 1997), are most appropriate. Specifically, this approach captures the interpersonal context of aggressive acts often observed within inpatient wards, with staff-patient interactions associated with treatment or maintenance of ward regime that are considered provocative or that threaten status. Table 6.3 outlines a number of psychosocial intervention strategies that can be used in response to various aggressor motivations. Take some time to consider whether any of these interventions might be helpful to use in response to dynamic risk behaviours exhibited by Maurice in Box 6.2.

Table 6.3 Psychosocial intervention strategies by aggressor motivation typology

Motive/ Classification	Intervention
Fear	✓ Address and modify any distorted beliefs ✓ Listen to patient concerns and develop safety plan ✓ Anger management programmes specifically to reduce anxiety and arousal ✓ Assertive and effective communication skills training ✓ Provision of validation, reassurance, and empathy
Frustration	✓ Clarify expectations of patient and staff member ✓ Relaxation strategies ✓ Explain why requests cannot be me ✓ Identify and use less provocative means of denying requests ✓ Assertive and effective communication skills training
Instrumental	✓ Avoid delivering tangible goods following the expression of aggression ✓ Teach adaptive means of securing goods or acquiring situational goal ✓ Address and modify any distorted beliefs about use of aggression ✓ Clarify legitimate channels for getting needs met ✓ Identify and reinforce effective methods of securing goods
Pleasure	✓ Explore alternative channels for achieving pleasure ✓ Clarify the potential costs of aggressive behaviour ✓ Enforce consequences following displays of aggression ✓ Model prosocial behaviours via unit activities
Psychotic	✓ Explore delusions/hallucinations and create safety plan ✓ Establish alternative means of managing experiences ✓ Consider review of psychotropic medication ✓ Create advanced directives/management plan ✓ Consider environmental restrictions

Source: adapted from Daffern et al. (2007)

The strength of this approach is that it offers psychosocial interventions rather than physical restraint or pharmacological options for managing aggressive incidents that can be carried out by nursing staff. However, it is limited by its inability to provide a method of assessing imminent or short-term risk, and triage intensity of intervention required to prevent future aggressive acts.

Dynamic Appraisal and Management of Situational Aggression

There is an acute need to establish effective and feasible methods to predict, prevent, and manage aggressive incidents within psychiatric settings (Joint Commission, 2018). Any suitable tool should be able to offer the following: prospective guidance regarding likelihood of future aggression; the nature and function of an individual's likely aggression; a retrospective analysis of any aggressive incident; imminent, rather than chronic, predictive ability; and be readily communicable to an array of mental health professionals. Furthermore, many of the interventions (pharmacological, psychosocial, and environmental) outlined by Taylor (2018) that can be tailored to each level of DASA risk, are either pharmacological or environmentally coercive in nature. Thus, further supporting the findings of Kaunomäki et al. (2017) that DASA-informed interventions are rarely psychosocial in nature.

It is our contention that an adapted version of the DASA, which we have named the Dynamic Assessment and Management of Situational Aggression (DAMSA) and outlined below, may be capable of addressing the needs and issues outlined above. Our adapted version of the DASA attempts to provide an important bridge between established theories of aggression and imminent risk prediction, formulation models for forensic inpatient settings, as well as brief intervention strategies. Specifically, it is our contention that adapting the DASA assists healthcare professionals to reframe aggression predictors into meaningful clinical communications. This elevates a prediction assessment into a useful formulation tool that is more amenable to effective clinical prevention and intervention. It is our contention that without a theoretical understanding of the function of aggression and associated risk indicators, management is likely to remain focused on pharmacological and environmental interventions. Accordingly, we suggest that each behavioural risk indicator on the DASA is considered independently with respect to the social context using the following structured professional judgment considerations: (i) behaviour indicator (e.g. what were they doing/saying that constituted evidence for this item?); (ii) target focus (e.g. was there a target for this item, a person, object?); (iii) context or situation (e.g. in what situational context and time was this risk indicator observed?); (iv) possible classification or motive for behavioural indicator/s (e.g. fear, frustration, instrumental, pleasure, psychotic?); (v) if a significant aggressive act took place (e.g. verbal or physical aggression towards object, staff, patient, self); (vi) possible unmet need (e.g. what were/are they trying to achieve, obtain, convey?); and (vii) intervention plan (e.g. how can needs be addressed, met, or minimised). We recommend that these considerations be addressed (ideally in consultation with the patient) for individuals scoring 1 or more on the DASA.

These additional considerations aim to assist mental health professionals in developing a psychological understanding of the function of each patient's risk indicators. Ultimately, this will likely lead to more tailored psychosocial interventions. We suggest that the DAMSA can be used for the following purposes: (1) a prospective (imminent risk prediction) and retrospective (formulation) assessment; (2) a dynamic management planning instrument; and finally, (3) a staff supervision and reflection tool. Using the current case example introduced in Box 6.2, we will attempt to demonstrate the applicability of the DAMSA.

The DAMSA as a Prospective and Retrospective Risk Assessment

Using the DAMSA (Figure 6.1) for prospective (imminent risk prediction) and retrospective (formulation) purposes, let us once again return to the nursing note in Box 6.2 regarding Maurice, a young Black British male, residing on a forensic psychiatric unit. Although brief, there is enough information within this note to score the DAMSA. Maurice would have likely scored a 4 on the DASA (i.e. irritability, unwillingness to follow directions, easily angered when requests are denied, and negative attitudes), thus exceeding the threshold of 1 and requiring a more detailed analysis of his observed risk indicators.

Examining the behavioural indicators, likely targets, and social context associated with each item, provides a greater source of information regarding the possible nature and function of Maurice's presentation during the shift. We can see from Figure 6.1 that many of the DASA risk indictors involved Maurice pacing the corridors, avoiding his peers, and getting frustrated about not being able to use the unit telephone. In addition, it appeared that staff members were often the targets in the context of busy areas on the unit where other patients were present. This information can be used in collaboration with Maurice to further explore and understand these observations. Table 6.4 outlines a number of questions that can be used to elicit possible classifications or motivations for aggressive behaviours. Questions should be framed in a non-judgmental, neutral, and cautious way. For example, rather than asking "what were you fearful or frightened of?", which will most likely be met with resistance and denial, you might ask "what was concerning or bothering you at the time?"

This may confirm or uncover a number of possible unmet needs. For example, it may be revealed from discussions with Maurice that he was

	Irritability	Impulsivity	Unwillingness to follow directions	Sensitive to perceived provocation	Easily angered	Negative attitudes	Verbal threats
DASA ITEM PRESENT?	☐	☐	☐	☐	☐	☐	☐

IF 1+ ITEMS PRESENT OR AGGRESSIVE INCIDENT

BEHAVIORAL INDICATOR *What were they specifically doing or saying?*	Refusing to engage, not talking to peers, agitated	N/A	Trying to use broken telephone, not going to room	N/A	Became hostile and angry when told not to use phone	Making disparaging comments	N/A
TARGET OR FOCUS *Was target patient, self, staff, object?*	No clear target	N/A	Staff members	N/A	Staff members and other patients	Staff members	N/A
CONTEXT/ SITUATION *Where were they? What time of day?*	Throughout the shift, in, unit corridors and by telephone	N/A	In busy common areas, later in the afternoon	N/A	In busy common areas, later in the afternoon	In busy common areas, later in the afternoon	N/A

MEET AND FORMULATE

	Fear	Frustration	Instrumental	Pleasure	Psychotic

Figure 6.1 The DAMSA for imminent risk prediction and formulation

LIKELY CLASS/ MOTIVE *Fear, Frustration, Instrumental, Pleasure, Psychotic?*	☐		☐		☐	☐	☐
IF AGGRESSIVE ACT PRESENT* *Nature and harm caused?*	Verbal ☐	Physical Object ☐	Physical Staff ☐	Physical Patient ☐	Physical Self ☐		Severity of harm caused N/A
POSSIBLE UNMET NEEDS *What were/are they trying to achieve, obtain, convey?*		1. To be validated by staff members 2. To protect his physical safety on the unsettled unit 3. To make contact with lawyer to talk through upcoming court appearance 4. To convey anxiety about upcoming court appearance
	**DEVELOP BRIEF INTERVENTION PLAN**
INTERVENTION PLAN *How can needs be addressed, met, or minimised?*		1. Schedule meeting to discuss situation with key worker (fear/frustration) 2. Make a point of checking in with him periodically throughout shift (fear) 3. Ensure constant access to bedroom if feeling unsettled (fear) 4. Allow temporary use of nursing station phone (frustration) 5. Practice relaxation strategies (frustration)

Figure 6.1 (Cont.)

Table 6.4 Exploratory questions for eliciting likely motives for aggressor motivations

Motivations	Exploratory Questions
Fear	What was bothering you at the time? Were you feeling unsafe? How so? What were you feeling concerned about?
Frustration	Were you being stopped from doing or having something? Were your requests/needs being denied? Do you feel you were being ignored?
Instrumental	What do/did you hope your behaviour would achieve? What were the benefits of your behaviour? Were you feeling angry or irritable at the time?
Pleasure	What did the target of your anger learn? What were your feelings during/immediately following the incident? What was the best part of the altercation?

anxious and worried about his upcoming court appearance, could not get in contact with his lawyer, and felt unsettled and neglected due to the sudden influx of new patients on the unit. This information also suggests that fear and frustration were likely motivators for his current risk indicators. Accordingly, one may work with Maurice to create a brief intervention plan to address, meet, or minimise these unmet needs. As can be seen from Figure 6.1, all of these preventative interventions are psychosocial in nature and can be classified in accordance with the likely motivation. However, staff may also draw on unit-specific prevention strategies, as well as pharmacological options, if deemed necessary to manage risk.

The DAMSA as Dynamic Management Planning Instrument

In addition to using the DAMSA to prevent possible violence, it can also be used following an aggressive incident to develop a robust dynamic intervention/management plan. This purpose requires a more detailed examination of the incident, and a more in-depth interview with the aggressor. For the purpose of illustration, let us consider the aggressive incident outlined in Box 6.3.

Box 6.3 Aggressive incident

Maurice was asked by Serena, a relatively new female staff member, to stop pacing and knocking on the nursing station window. She told him he had to wait until the team meeting was finished before someone would come and speak to him. After another 10 minutes, he began banging on the window directly in front of Serena who was writing up an earlier incident. Serena came out of the nursing station, confronted him, and instructed him to return to his room immediately. Now very aggravated, he began to call her names. Maurice then attempted to bang on the window again and noticed her flinch. Maurice began moving towards Serena, invading her personal space, and looking down at her. Despite being intimidated, Serena refused to move, and began to laugh nervously. Embarrassed and taken aback by her reaction, Maurice spontaneously pushed her against the window and made degrading comments about her physical appearance. She immediately cowered. Satisfied with her response, Maurice told her not to move. Serena remained where she was and began shaking. Maurice started to laugh at her. Serena pressed her personal alarm signalling the emergency response team and ran back inside the nursing station. "You are useless. You can't do anything yourself, can you?" shouted Maurice before he was restrained and moved to the seclusion room.

Closer examination of this incident actually reveals a chain or collection of smaller aggressive incidents that escalated between Maurice and Serena. Furthermore, these aggressive incidents reflect a mix of motivations. Maurice was displaying clear signs of irritability (e.g. pacing and banging on the nursing station window), impulsiveness (e.g. pushing Serena against the window), unwillingness to follow directions (e.g. waiting until a team meeting was over, and returning to his room), being easily angered when requests were denied (e.g. unable to speak to anyone, not able to use phone), and negative attitudes (e.g. calling Serena useless, not following her demands). A retrospective analysis would not only provide details about Maurice's risk prediction profile (or signature early warning signs) but also indicate the possible motivations and subsequent unmet needs that may have culminated in this aggressive incident. Meeting with the patient and the target of the aggressive incident will also provide further important details about the event.

Box 6.4 provides useful information from the perspective of Maurice. It not only reveals the possible motives behind the aggressive incident but discloses the interpersonal sequence of possible motivators. To begin, one might speculate that fear was initially driving his risk indicators. For example, "I needed someone to call my attorney about my court hearing", "she never showed up", and "that was not like her". There were also new patients on the ward causing disruption and causing him to feel unsettled. Next in the sequence is frustration which appears to take over as the driving force. For example, "I asked Serena twice", "she kept saying 'give me a minute'", "she just kept ignoring me", and "she tells me to go to my room". We can also speculate a number of blocked goals; not being able to contact his lawyer, obtain immediate assistance from staff, remain at the nursing station, or be taken seriously. In the next step, it is unclear whether fear or frustration, or a combination of both, triggered Maurice's most significant aggressive behaviour (i.e. pushing Serena). This act of aggression did not appear planned or premeditated as it took place impulsively ("I wanted to explode", "I pushed her", "I didn't mean to") during a period of high arousal. However, one might speculate that Maurice initially felt invalidated, disrespected, vulnerable, and/or shamed by Serena's nervous laughter ("she wasn't taking me seriously at all and then laughed at me", "so bloody disrespectful"). One may speculate that some behavioural reinforcement and script development might have occurred following his observation that "only then did she take me seriously". There is also evidence that pleasure was a further contributing motivator as he observed Serena appear submissive, frightened, and vulnerable as he continued to stare down at her. Accordingly, as illustrated in Figure 6.2, fear, frustration, and pleasure are marked as the most likely motivators. Furthermore, we can indicate the nature of, and likely harm caused by, this aggressive incident on the primary target (Serena) – psychological harm, fear, and intimidation.

136 Matt Bruce & Veronica Rosenberger

	MEET AND FORMULATE				

LIKELY CLASS/ MOTIVE Fear, Frustration, Instrumental, Pleasure, Psychotic?	Fear ☐	Frustration ☐	Instrumental ☐	Pleasure ☐	Psychotic ☐
IF AGGRESSIVE ACT PRESENT* Nature and harm caused?	Verbal ☐	Physical Object ☐	Physical Staff ☐	Physical Patient ☐	Severity of harm caused Psychological harm Fear and intimidation
POSSIBLE UNMET NEEDS What were/are they trying to achieve, obtain, convey?	1. To make contact with lawyer to talk through upcoming court appearance 2. To achieve control over an unpredictable situation 3. To feel respected and understood by staff 4. To express an urgent need to express wishes				

	DEVELOP INTERVENTION PLAN

INTERVENTION PLAN How can needs be addressed, met, or minimised?	1. Address and modify any distorted beliefs (fear) 2. Listen to Maurice's concerns and develop safety plan (fear) 3. Explain to Maurice why requests cannot be met (frustration) 4. Assertive and effective communication skills training (frustration) 5. Model prosocial behaviours to Maurice via role modelling and unit activities (pleasure) 6. Relaxation strategies (frustration)

* To be completed only following aggressive incident

Figure 6.2 The DAMSA for dynamic management planning

Box 6.4 Maurice's reflections of the incident

I needed someone to call my attorney about my court appearance. I was supposed to see her that afternoon, but she never showed up. That was not like her. The unit social worker is off on sick leave so I couldn't ask her. New patients kept causing issues and using up all the nurse's time. I had asked Serena twice and she kept saying, "give me a minute". After 30 minutes, I tapped on the window, but she just kept ignoring me like I wasn't worth shit. It really pissed me off. She had spent over an hour with a new patient, and I just needed 5 minutes. Then she tells me to go to my room like I'm some kid! She wasn't taking me seriously at all and then she laughed at me! So bloody disrespectful. I wanted to explode. I pushed her. I didn't mean to. Only then did she take me seriously. I wanted to watch her squirm a bit. She deserved that. I would have gone to my room like she asked in the end. She called the response team out of spite! Now what happened will be recorded as a serious incident which will not look good when I next appear in court.

Given that Maurice's aggression was likely driven by fear, frustration, and pleasure, the following brief psychosocial interventions for dynamic management planning were recommended (see Figure 6.2). Address and modify any distorted beliefs (Renwick et al., 1997), listen to his concerns and develop safety plan, explain to Maurice why requests cannot be met (Daffern et al., 2007), provide him with assertive and effective communication skills training (Winogron et al., 2001), model prosocial behaviours to him via role modelling and unit activities (Braham et al., 2008), and provide relaxation strategies (Novaco, 1977). Please refer to Table 6.3 for a list of possible psychosocial intervention strategies for both patient and mental health professional to engage in. As it is our contention that all aggressive behaviour is inherently linked to mental health staff and environmental characteristics (Johnson, 2004; Cutcliffe & Riahi, 2018), we argue that all psychosocial interventions should be interpersonal in nature.

The DAMSA for Staff Supervision and Reflection

As the genesis of interpersonal aggression is inherently rooted within social systems and relationships, the DAMSA can also be used as a tool to advance understanding and insight into the actions of staff members. For example, do you think that Maurice could have perceived Serena's words or actions as aggressive, controlling, forceful, or threatening? Following an act of aggression between inpatients and/or staff members, it is important to examine the experience of both parties. This will allow a more useful and accurate understanding of the nature and function of the aggressive incident. Study Serena's experience (Box 6.5). Compare and contrast her perspective of the incident with that of Maurice's (Box 6.4). Do you think any of her thoughts or behaviours could have been interpreted as aggressive?

Box 6.5 Serena's experience

It was the end of my 12-hour shift, and I was exhausted. I was also in a bad mood because my unit manager had just refused my annual leave request. There were three new admissions and we had to cancel groups today due to staff shortages. Maurice had been so demanding all day. Maurice refused to follow my directions. He just wasn't letting up and kept tapping on the window right in front of me. He was just trying to wind me up and I couldn't concentrate on what I was doing. So, I got up, confronted him on the unit floor, and as per unit protocol, I instructed him to return to his room. This seemed to just anger him more and that's when he moved towards me shouting and swearing. Then he pushed me. I felt scared and vulnerable. I decided that he was not going to get away with thinking that he could intimidate me. I had had enough. I wanted to teach him a lesson, so I called for the emergency response team. I thought, he'll think twice before trying that with me again.

There appears to have been important pieces of information that each person neglected to consider, which directly influenced how they perceived and responded to the actions of the other. While Serena appeared unaware of Maurice's escalating anxiety over the influx of new patients and the whereabouts of his lawyer, he was oblivious to her increasing levels of stress, exhaustion, and frustration regarding her annual leave being denied and staff shortages. This mutual deficit in understanding the needs of each person ultimately led Serena to respond to Maurice's escalating aggressive behaviours by calling the crisis response team. This resulted in an emergency environmental intervention involving restraints and seclusion.

Through the process of supervision, Serena's supervisor could ask her to complete or review Maurice's DAMSA assessment. However, to help Serena generate greater levels of insight and self-awareness, she could also be assisted in completing the DAMSA as a way to examine her own behaviours during the incident. She may likely endorse a number of dynamic behavioural items that were present for her at the time (e.g. irritability, sensitivity to perceived provocation, negative attitudes). This, in turn, may lead to a useful reflection and identification of various behavioural indicators (e.g. being unable to concentrate on work, accusing patients of deliberately trying to provoke her, and selecting punitive interventions), the target or focus (e.g. Maurice), and situational context (e.g. in the nursing station at the end of a busy, stressful, frustrating shift). Furthermore, Serena may identify fear, frustration, and instrumentality as the likely motives behind her decision to call the emergency response team in this particular situation.

It is important to note that the goal of completing the DAMSA for both parties involved in the aggressive incident is not to apportion blame or responsibility, but instead help staff reflect on the psychosocial interpersonal context of aggressive incidents. In some circumstances, aggression may be triggered exclusively and obviously by internal stimuli that is relatively unrelated to the situational or interpersonal context (e.g. psychotic motives). However, most of the time it takes place within a social context which must be examined in order to assist future prevention. Indeed, Cutcliffe and Riahi (2013) have argued that the role of staff in violent and aggressive incidents is often ignored, despite growing evidence that suggests clinicians have a significant influence. In their review of the literature, they found that staff who possessed, and were conversant with, communication and interpersonal skills that conveyed a sense of respect, capacity to listen, and ability to lower the relational "temperature", were less likely to report involvement in aggressive incidents.

In sum, regular use and review of completed DAMSA assessments of both parties involved in aggressive incidents may provide multiple benefits: (i) improved staff understanding of inpatients' risk behaviours, related interpersonal context, possible motives for aggression, and likely unmet needs; (ii) development of effective and tailored risk management plans to minimise future aggressive incidents; and (iii) increased understanding by staff members of the role their attitudes and behaviours may play in the commission of aggressive acts.

Conclusion

While the construct of aggression encompasses an immeasurable range of biopsychosocial phenomena, it is, by definition, an inherently social artefact. Aggression can be understood as an important social tool for the management of interpersonal relationships and interactions for the purposes of self-preservation, demand avoidance, gratification, and goal acquisition. Through the use of case examples and practice exercises, this chapter provided an overview of social psychological theory, assessment, prediction, and management of aggression in forensic mental health settings. It also outlined the development of an adapted assessment tool (DAMSA) which attempts to identify and conceptualise the dynamic and fluctuating nature of acute mental illness in forensic psychiatric settings, to protect patients and staff, while creating a safer and psychologically informed environment. The tool recognises that aggression in clinical settings cannot be separated from interactions with staff and the surrounding environment (Johnson, 2004; Cutcliffe & Riahi, 2018). Accordingly, the DAMSA developed in this chapter offers theoretically grounded and empirically derived guidance for predicting imminent risk and conceptualising aggressive outbursts. It further assists in the development of tailored and evidence-based social psychological interventions in forensic mental health settings. It is hoped that the use of the DAMSA to assess behavioural motivation and associated risk indicators for all individuals interacting within clinical environments, will lead to collaborative interventions that decrease the likelihood of institutionally based aggressive behaviour.

References

Almvik, R., Woods, P., & Rasmussen, K. (2000). The Brøset Violence Checklist: sensitivity, specificity, and interrater reliability. *Journal of Interpersonal Violence*, 15 (12), 1284–1296.

Anderson, C. A., & Bushman, B. J. (2002). Human aggression. *Annual Review of Psychology*, 53, 27–51.

Bandura, A. (1973). *Aggression: A social learning analysis*. Prentice-Hall.

Baron, R. A., & Richardson, D. R. (1994). *Human aggression*. Springer Science & Business Media.

Barry-Walsh, J., Daffern, M., Duncan, S., & Ogloff, J. (2009). The prediction of imminent aggression in patients with mental illness and/or intellectual disability using the Dynamic Appraisal of Situational Aggression instrument. *Australasian Psychiatry*, 17(6), 493–496.

Berkowitz, L. (1962). *Aggression: A social psychological analysis*. McGraw-Hill.

Berkowitz, L. (1989). Frustration-aggression hypothesis: examination and reformulation. *Psychological Bulletin*, 106(1), 59–73.

Berkowitz, L. (1990). On the formation and regulation of anger and aggression: A cognitive neoassociationistic analysis. *American Psychologist*, 45(4), 494–503.

Berkowitz, L. (1993). Pain and aggression: Some findings and implications. *Motivation and Emotion*, 17(3), 277–293.

Bjørkly, S., Hartvig, P., Heggen, F. A., Brauer, H., & Moger, T. A. (2009). Development of a brief screen for violence risk (V-RISK-10) in acute and general psychiatry: An introduction with emphasis on findings from a naturalistic test of interrater reliability. *European Psychiatry*, 24(6), 388–394.

Borduin, C. M. (1999). Multisystemic treatment of criminality and violence in adolescents. *Journal of the American Academy of Child and Adolescent Psychiatry*, 38(3), 242–249.

Braham, L., Jones, D., & Hollin, C. R. (2008). The Violent Offender Treatment Program (VOTP): development of a treatment program for violent patients in a high security psychiatric hospital. *International Journal of Forensic Mental Health*, 7(2), 157–172.

Breuer, J., & Elson, M. (2017). Frustration–aggression theory. In P. Sturmer (ed.), *The Wiley handbook of violence and aggression*, 1–12. John Wiley.

Bruce, M. (2014). Violent Offending: Assessment and Treatment in the Community. *Forensic Practice in the Community*, 5, 101–124.

Bushman, B. J., & Anderson, C. A. (2001). Is it time to pull the plug on hostile versus instrumental aggression dichotomy? *Psychological Review*, 108(1), 273–279.

Chichinadze, K., Chichinadze, N., & Lazarashvili, A. (2011). Hormonal and neurochemical mechanisms of aggression and a new classification of aggressive behavior. *Aggression and Violent Behavior*, 16(6), 461–471.

Cutcliffe, J. R., & Riahi, S. (2013). Systemic perspective of violence and aggression in mental health care: towards a more comprehensive understanding and conceptualization: part 2. *International Journal of Mental Health Nursing*, 22(6), 568–578.

Cutcliffe, J. R., & Riahi, S. (2018). A systematic perspective of violence and aggression inmental health care: Toward a more comprehensive understanding and conceptualization. In J. C. Santos & J. R. Cutcliffe (eds), *European psychiatric/mental health nursing in the 21st century: A person-centred evidence-based approach* (pp. 453–477). Springer.

Daffern, M., Howells, K., & Ogloff, J. (2007). What's the point? Towards a methodology for assessing the function of psychiatric inpatient aggression. *Behaviour Research and Therapy*, 45(1), 101–111.

de Mooij, L. D., Kikkert, M., Lommerse, N. M., Peen, J., Meijwaard, S. C., Theunissen, J., Duurkoop, P. W., Goudriaan, A. E., Van, H. L., Beekman, A. T. F., & Dekker, J. J. (2015). Victimisation in adults with severe mental illness: prevalence and risk factors. *The British Journal of Psychiatry*, 207(6), 515–522.

Dodge, K. A., Pettit, G. S., & Bates, J. E. (1994). Socialization mediators of the relation between socioeconomic status and child conduct problems. *Child Development*, 65(2), 649–665.

Dodge, K. A., Pettit, G. S., McClaskey, C. L., Brown, M. M., & Gottman, J. M. (1986). Social competence in children. *Monographs of the Society for Research in Child Development*, 51, 1–85.

Dollard, J., Miller, N. E., Doob, L. W., Mowrer, O. H., & Sears, R. R. (1939). *Frustration and aggression*. Yale University Press.

Douglas, K. S., Hart, S. D., Webster, C. D., Belfrage, H., Guy, L. S., & Wilson, C. M. (2014). Historical-clinical-risk management-20, version 3 (HCR-20V3): development and overview. *International Journal of Forensic Mental Health*, 13(2), 93–108.

Elson, M., & Ferguson, C. J. (2014). Does doing media violence research make one aggressive? The ideological rigidity of social cognitive theories of media violence and response to Bushman and Huesmann (2013), and Krahé (2013). *European Psychologist*, 19, 68–75.

Felson, R. B., & Tedeschi, J. T. (1993). *Aggression and violence: Social interactionist perspectives*. American Psychological Association.

Gendreau, P. L., & Archer, J. (2005). Subtypes of aggression in humans and animals. In R. E. Tremblay, W. W. Hartup, & J. Archer (eds), *Developmental origins of aggression* (pp. 25–46). The Guilford Press.

Gilbert, F., & Daffern, M. (2010). Integrating contemporary aggression theory with violent offender treatment: How thoroughly do interventions target violent behavior? *Aggression and Violent Behavior*, 15(3), 167–180.

Goldstein, A. P., Nensén, R., Daleflod, B., & Kalt, M. (Eds.). (2004). *New perspectives on aggression replacement training: Practice, research, and application*. Wiley.

Griffith, J. J., Daffern, M., & Godber, T. (2013). Examination of the predictive validity of the dynamic appraisal of situational aggression in two mental health units. *International Journal of Mental Health Nursing*, 22(6), 485–492.

Hatcher, R. M., Palmer, E. J., McGuire, J., Hounsome, J. C., Bilby, C. A., & Hollin, C. R. (2008). Aggression replacement training with adult male offenders within community settings: a reconviction analysis. *The Journal of Forensic Psychiatry & Psychology*, 19(4), 517–532.

Henggeler, S. W., Schoenwald, S. K., Borduin, C. M., Rowland, M. D., & Cunningham, P. B. (2009). *Multisystemic therapy for antisocial behavior in children and adolescents* (2nd edition). Guilford Press.

Hipp, K., Repo-Tiihonen, E., Kuosmanen, L., Katajisto, J., & Kangasniemi, M. (2020). PRN Medication Events in a Forensic Psychiatric Hospital: A Document Analysis of the Prevalence and Reasons. *International Journal of Forensic Mental Health*, 19(4), 329–340.

Howells, K., & Hollin, C. R. (1989). *Clinical approaches to violence*. Wiley.

Huesmann, L. R. (1988). An information processing model for the development of aggression. *Aggressive Behavior*, 14(1), 13–24.

Huesmann, L. R. (1998). The role of social information processing and cognitive schema in the acquisition and maintenance of habitual aggressive behavior. In R. G. Geen & E. Donnerstein (eds), *Human aggression: Theories, research, and implications for social policy* (pp. 73–109). Academic Press.

Johnson, M. E. (2004). Violence on inpatient psychiatric units: State of the science. *Journal of the American Psychiatric Nurses Association*, 10, 113–121.

Joint Commission. (2018). Physical and verbal violence against health care workers. Retrieved from www.jointcommission.org/-/media/tjc/idev-imports/topics-assets/workplace-violence-prevention-implementing-strategies-for-safer-healthcare-organizations/sea_59_workplace_violence_4_13_18_finalpdf.pdf.

Jolliffe, D., & Farrington, D. P. (2007). *A systematic review of the national and international evidence on the effectiveness of interventions with violent offenders*. Ministry of Justice.

Jones, M. (ed.). (2013). *Social psychiatry: A study of therapeutic communities* (vol. 4). Routledge.

Kaunomäki, J., Jokela, M., Kontio, R., Laiho, T., Sailas, E., & Lindberg, N. (2017). Interventions following a high violence risk assessment score: a naturalistic study on a Finnish psychiatric admission ward. *BMC Health Services Research*, 17(1), 1–8.

Kempes, M., Matthys, W., De Vries, H., & Van Engeland, H. (2005). Reactive and proactive aggression in children A review of theory, findings and the relevance for child and adolescent psychiatry. *European Child & Adolescent Psychiatry*, 14(1), 11–19.

Lantta, T., Anttila, M., Kontio, R., Adams, C. E., & Välimäki, M. (2016). Violent events, ward climate and ideas for violence prevention among nurses in psychiatric wards: a focus group study. *International Journal of Mental Health Systems*, 10(1), 1–10.

Malivert, M., Fatséas, M., Denis, C., Langlois, E., & Auriacombe, M. (2012). Effectiveness of Therapeutic Communities: A Systematic Review. *European Addiction Research*, 18(1), 1–11. doi:10.2307/26790581.

McDermott, B. E., & Holoyda, B. J. (2014). Assessment of aggression in inpatient settings. *CNS Spectrums*, 19(5), 425–431.

McDermott, B. E., Quanbeck, C. D., Busse, D., Yastro, K., & Scott, C. L. (2008). The accuracy of risk assessment instruments in the prediction of impulsive versus predatory aggression. *Behavioral Sciences & the Law*, 26(6), 759–777.

Megargee, E. I. (1982). Psychological assessment in jails: Implementation of the standards recommended by the National Advisory Commission on Criminal Justice Standards. In NCJRS (ed.), *Mental health services in local jails*, 100–125. NCJRS.

Megargee, E. I. (2009). Understanding and assessing aggression and violence. In J. N. Butcher (ed.), *Oxford handbook of personality assessment*, 542–566. Oxford University Press.

Merk, W., Orobio de Castro, B., Koops, W., & Matthys, W. (2005). The distinction between reactive and proactive aggression: Utility for theory, diagnosis and treatment? *European Journal of Developmental Psychology*, 2(2), 197–220.

Monahan, J., Steadman, H. J., Robbins, P. C., Appelbaum, P., Banks, S., Grisso, T., Heilbrun, K., Mulvey, E. P., Roth, L., & Silver, E. (2005). An actuarial model of violence risk assessment for persons with mental disorders. *Psychiatric Services*, 56(7), 810–815.

National Institute for Health and Care Excellence. (2015). *Violence and aggression: short-term management in mental health, health and community settings*. NICE Guideline no. 10. Retrieved from www.nice.org.uk/guidance/ng10.

Nolan, K. A., Czobor, P., Roy, B. B., Platt, M. M., Shope, C. B., Citrome, L. L., & Volavka, J. (2003). Characteristics of assaultive behavior among psychiatric inpatients. *Psychiatric Services*, 54(7), 1012–1016.

Novaco, R. W. (1975). *Anger control: The development and evaluation of an experimental treatment*. Lexington.

Novaco, R. W. (1977). Stress inoculation: A cognitive therapy for anger and its application to a case of depression. *Journal of Consulting and Clinical Psychology*, 45(4), 600.

Novaco, R. W. (2013). Reducing anger-related offending: What works. In L. A. Craig, L. Dixon, & T. A. Gannon (eds), *What works in offender rehabilitation: An evidence-based approach to assessment and treatment* (pp. 211–236). Wiley Blackwell.

Nussbaum, D., Saint-Cyr, J., & Bell, E. (1997). A biologically derived, psychometric model for understanding, predicting and treating tendencies toward future violence. *American Journal of Forensic Psychiatry*, 18, 35–50.

Ogloff, J. R. P., & Daffern, M. (2003). *The assessment of inpatient aggression at the Thomas Embling Hospital: Toward the dynamic appraisal of inpatient aggression*. Victorian Institute of Forensic Mental Health Fourth Annual Research Report to Council. Forensicare.

Ogloff, J. R., & Daffern, M. (2006). The dynamic appraisal of situational aggression: an instrument to assess risk for imminent aggression in psychiatric inpatients. *Behavioral Sciences & the Law*, 24(6), 799–813.

Phenix, A., Helmus, L., & Hanson, R. K. (2016). *Static-99R and Static-2002R: Evaluators' workbook*. Public Safety Canada. Retrieved from www.static99.org/pdfdocs/Evaluators_Workbook_2016-10-19.pdf.

Polaschek, D. L., & Reynolds, N. (2004). Assessment and treatment: Violent offenders. In C. R. Hollin (ed.), *The essential handbook of offender assessment and treatment*, 201–218. John Wiley.

Quanbeck, C. D., McDermott, B. E., Lam, J., Eisenstark, H., Sokolov, G., & Scott, C. L. (2007). Categorization of aggressive acts committed by chronically assaultive state hospital patients. *Psychiatric Services*, 58(4), 521–528.

Ramesh, T., Igoumenou, A., Montes, M. V., & Fazel, S. (2018). Use of risk assessment instruments to predict violence in forensic psychiatric hospitals: a systematic review and meta-analysis. *European Psychiatry*, 52, 47–53.

Renwick, S. J., Black, L., Ramm, M., & Novaco, R. W. (1997). Anger treatment with forensic hospital patients. *Legal and Criminological Psychology*, 2(1), 103–116.

Skinner, B. F. (1963). Operant behavior. *American Psychologist*, 18(8), 503.

Taylor, K. (2018). *Crisis prevention and management: Best practices from inpatient psychiatric units*. Johns Hopkins Medicine Bayview Medical Centre.

Urheim, R., Rypdal, K., Melkevik, O., Hoff, H. A., Mykletun, A., & Palmstierna, T. (2014). Motivational dimensions of inpatient aggression. *Criminal Behaviour and Mental Health*, 24(2), 141–150.

Vandevelde, S., Broekaert, E., Yates, R., & Kooyman, M. (2004). The development of the therapeutic community in correctional establishments: A comparative retrospective account of the 'democratic' Maxwell Jones TC and the hierarchical concept-based TC in prison. *International Journal of Social Psychiatry*, 50(1), 66–79.

Vitacco, M. J., Van Rybroek, G. J., Rogstad, J. E., Yahr, L. E., Tomony, J. D., & Saewert, E. (2009). Predicting short-term institutional aggression in forensic patients: A multi-trait method for understanding subtypes of aggression. *Law and Human Behavior*, 33(4), 308–319.

Walsh, E., Moran, P., Scott, C., McKenzie, K., Burns, T. O. M., Creed, F., Tyrer, P., Murray, R. M., & Fahy, T. (2003). Prevalence of violent victimisation in severe mental illness. *The British Journal of Psychiatry*, 183(3), 233–238.

Warburton, W. A., & Anderson, C. A. (2015). Aggression, social psychology of. *International encyclopedia of the social & behavioral sciences*, 373–380. Elsevier.

Webster, C. D., Douglas, K. S., Eaves, D., & Hart, S. D. (1997). Assessing risk of violence to others. In C. D. Webster & M. A. Jackson (eds), *Impulsivity: Theory, assessment, and treatment*, 251–277. Guilford Press.

Winogron, W., van Dieten, M., & Grisim, V. (2001). *Controlling Anger and Learning to Manage It – Effective Relapse-Prevention Program (CALMER)*. Multi-Health Systems.

Zillmann, D. (1971). Excitation transfer in communication-mediated aggressive behavior. *Journal of Experimental Social Psychology*, 7(4), 419–434.

7 Group Formation and Behaviour

Derval Ambrose and Tania Tancred

> We must live in groups; other people are like nutrients for us, and are absolutely essential for our survival.
>
> (W. Gaylin, personal communication)

Introduction

The majority of the rehabilitation, care, and control of those who come into contact with criminal justice systems occurs in group situations. For example, most of the rehabilitation of forensic clients in the United Kingdom (UK), Canada, and Scandinavia, is *group-based*, and often includes both offence-specific (i.e. treating the relevant psychological, behavioural, and lifestyle factors directly related to the offending behaviour) and task-specific group work (e.g. occupational therapy or attendance at therapeutic community meetings). The formulation of risk and need, and decisions about release or discharge from detention into the community, also occurs in group contexts. Additionally, group membership and group processes exist as intrinsic parts of the everyday life of staff in forensic settings- be that involvement with patient review meetings in secure health settings, or risk formulation meetings in prison and probation contexts.

Forensic clients enter the criminal justice system (CJS) with varying experiences of being part of a group, and a range of skills related to coping with group interactions. Previous experiences of group situations may have been very negative for much of this population- not least due to the 'imported vulnerability' that they bring with them into the system (Harvey, 2011). For example, individuals in forensic settings are more likely to have experienced care during childhood, been excluded from academic institutions, and have other adverse childhood experiences (ACES[1]) than non-forensic populations (Messina & Schepps, 2021). These traumatic experiences may well have influenced how individuals experience and manage group settings. Additionally, multiple experiences of being excluded or discriminated against can hinder an individual's chance of social inclusion by affecting, their psychological well-being, sense of agency, and access to opportunities (United Nations, 2018). Indeed, long-term exposure to discrimination can result in an internalisation of this prejudice and subsequent low self-esteem, feelings of shame, distress, and poor physical health outcomes (ibid.). This is very likely to have an impact on how individuals learn to manage and cope in group situations. For example, feeling a sense of isolation and exclusion while growing up may well have contributed to the development of a heightened sense of hypervigilance, or avoidant or disorganised attachment style. These coping styles are then likely to have a significant impact on how an individual experiences a group setting- something that professionals should, arguably, bear in mind during the early stages of an individual entering a criminal justice system. This is particularly relevant for individuals who are commencing a prison sentence, as they will typically have some form of group intervention included in their sentence plan.

Secure settings can often be characterised by an overwhelming variety of social relations and group activity, as well as protracted periods of isolation, or limited social interaction. In turn, individual autonomy over group membership can be very limited and, instead, determined by those in positions of power and authority. On admission to a secure setting, an individual is quickly faced with multiple group dynamics and group membership decisions. The following vignette demonstrates some of the likely group membership options that may present themselves to someone who is very early on in their prison experience.

> Rashid has received a life sentence for the rape and attempted murder of his female partner. He has started his sentence in a local category B prison, and spent his first night in a shared cell with another prisoner, John, who is serving four years for sexually assaulting a child.
>
> Rashid grew up living with his mother, who had ongoing mental health difficulties. Throughout his childhood, he spent periods of time in the care of the local authority when his mother was too unwell to take care of him. Rashid was sexually assaulted in one care home when he was eight years old. In school, he was a withdrawn and socially isolated child who, as a teenager, then got involved with a 'bad' crowd. During his teenage years, he was a member of a local gang and served several short sentences in young offender institutes (YOIs) for offences ranging from burglary and drug possession to car theft. Rashid regularly truanted from school, and left permanently without any qualifications when he was 15 years old.
>
> Within the first few weeks of his current sentence, Rashid is faced with the challenge of having to navigate different, and often competing, social groups. Some of the group membership tasks that face Rashid will feel coercive, while others seem to present themselves with elements of choice. For example, he may need to make decisions about the membership of the following groups:

- Early on, Rashid is approached by another life sentence prisoner who suggests that he would be wise to only socialise with other lifers.
- While out of his cell, during association time, Rashid is approached by another prisoner who asks him if he is a child sex offender like his cellmate. To prove that he is not like John, Rashid is encouraged to act aggressively towards him and distance himself from John as much as possible.
- Rashid's personal officer has suggested that Rashid needs to quickly decide what work or education he wants to get involved in to avoid remaining on a basic regime where he will not be able to progress towards having certain prison privileges (e.g. in-cell TV).
- Rashid's offender supervisor has suggested that he is placed on the waiting list to complete group work focused on addressing his risk of sexual offending.
- Rashid is offered alcohol and 'Spice' (new psychoactive substance) on day three of his stay. It is suggested that he needs to be friendly with a gang that operates on the wing.
- Rashid is approached by another prisoner who suggests that he attends prayer services.

These are just some examples of the types of group membership tasks that someone could face during the first few weeks of their custodial sentence. In Rashid's case, he must now quickly decide if he wants to comply with the system or protest against it. Additionally, he must decide which, if any, social groupings he wants to align himself with. Some of the group membership tasks presented to Rashid are likely to feel threatening and stressful. Also due to Rashid's early life experiences he may have adopted an avoidant coping style, which will heighten his internal stress when faced with an abundance of group membership tasks. The decisions Rashid makes about group membership will likely have a significant impact on how well he progresses through his sentence, and how bearable it is for him.

We will now explore some key social psychological theories from the perspective of groupwork during a period of detention and professional group decision-making at the point of exit from detention.

Groupwork during Detention

The Delivery of Accredited Behaviour Programmes

As seen in the case of Rashid, from the moment an individual begins a period of detention, and indeed throughout the duration of their sentence, they are faced with multiple group decisions/tasks that they must learn how to navigate. Over the past three decades, rehabilitation of forensic clients has become increasingly focused on the delivery of groupwork programmes. In

England and Wales, for example, accredited offending behaviour programmes were introduced into His Majesty's Prison Service in the early 1990s, and have since grown in volume, both in terms of delivery rate and variety (Falshaw et al., 2004). Some of these group interventions, particularly those relating to sexual offending, have largely been experimental in nature, and impact research has started to become more available in recent years (Towl & Crighton, 2016). The accreditation process was designed to ensure that offending behaviour groups were in line with the 'What Works?' effective practice literature, and delivered in a consistent manner across various sites (Andrews et al., 1990; Garrett, 1985; Izzo & Ross, 1990; Lipsey, 1992). In the UK, this process is overseen by the Correctional Services Accreditation Panel (CSAP). However, accessing details around how this panel operates exactly is difficult to establish (Towl & Podmore, 2019). During 2014/15, there were 8,523 accredited programme starts[2] delivered in prisons in the UK, at a rate of 11.9 starts per 100 offenders (Ministry of Justice, 2015).

The Probation Service in England and Wales similarly adopted accredited group programmes in the early 2000s, and introduced a wide variety of programmes, including cognitive skills groups (Ambrose, 2005) and specific offending behaviour groups, such as Sexual Offender group initiatives (Williams, 2005). During 2014/15, there were 14,023 accredited programme starts at a rate of 9.6 starts per 100 offenders (Ministry of Justice, 2015).

In Northern America, there has also been a long history of delivering group work in prison. For example, the Violence Reduction Programme (Wong & Gordon, 2013) and the Reasoning and Rehabilitation Programme (R&R; Ross et al., 1986) both originated in Canada, but are run in correctional and health facilities across the world. Similarly, cognitive behavioural groups are widely used in the United States (US), including programmes such as Thinking for a Change (TFAC) and the gender responsive Moving On initiative (Duwe, 2017; Van Dieten & MacKenna, 2001).

As a result of these developments in forensic rehabilitation, those who come into contact with the CJS will often be expected to complete various group programmes as identified by those involved in their care and rehabilitation. For example, offender managers[3] in the UK will devise a sentence plan that will often detail various group programmes that an individual must complete as part of their sentence. If the individual is serving a sentence that is subject to a parole board review before release, they will be expected to complete these programmes before they are likely to be considered successful at a parole board hearing. Despite the arguably coercive nature of this, in that release from custody or maintenance of probation conditions can rest upon groupwork completion, there are significant attrition rates involved in the delivery of groups to forensic populations. In community settings in the UK, it is estimated that between one third and a half of all group programme starters fail to complete (National Offender Management Service, 2005). While the non-completion rate in prisons is lower than that of community settings, there are still significant levels of group attrition; estimated at 9 per cent for adults and 14 per cent for young people who have offended (Cann et al., 2003). In addition to the obvious economic concern this presents, there is also increasing evidence that those who start treatment but do not complete are at a higher risk of re-offending (McMurran & Theodosi, 2007).

One of the principal models to emerge from the 'What Works?' literature is that of Risk-Need-Responsivity (RNR; Andrews & Bonta, 2003; Andrews & Dowden, 2006; Andrews et al., 1990, 2006; Gendreau & Andrews, 1990). This relates to three principles that are proposed to inform interventions with people who offend, in order to maximise effective outcomes (Andrews & Bonta, 2003). In summary, the *risk principle* refers to the idea that targeting treatment should be a priority for those at the highest risk of re-offending, and the extensiveness of treatment should be related to the risk levels presented. The *needs principle* recommends that treatment should directly target, and aim to change, areas that have been empirically demonstrated to increase the risk of future re-offending. This is in reference to so-called 'criminogenic' factors that have been demonstrated to result in the greatest reduction of re-offending. This is an arguably narrow focus and pays little to no attention to other important factors such as traumatic histories, distress, and low self-esteem.

The third *principle of responsivity* is most relevant to the discussion here. Responsivity is said to have two elements: specific and general (Bonta & Andrews, 2007). Specific responsivity refers to the idea that treatment should be adjusted to meet the needs of the individual. For example, adjusting interventions due to the participant's cultural background, intellectual capacity, or educational history. In essence, this aspect of responsivity demands a certain degree of flexibility in treatment (Marshall et al., 2013). The idea of general responsivity refers to the use of a broad range of cognitive social learning methods to influence behaviour. However, the conceptualisation of general responsivity appears to vary in the literature, with some claims suggesting that cognitive behavioural approaches are responsive because it has some of the best outcome research (ibid.). While this appears limited and off kilter with the idea of being responsive to need, it does, in fact, form the basis of most group interventions in forensic settings.

The literature strongly supports the idea that the *way* in which, and by whom, psychological therapy is delivered and supervised are the most critical aspects of implementation in relation to its effectiveness (Gannon et al., 2019). In addition to the influence of the RNR models interpretation, the operationalisation of this has resulted in a restricted menu of accredited programmes that place a heavy emphasis on consistent procedures. In turn, this has led to the increasing manualisation of rehabilitation group delivery (Marshall, 2009). Group interventions delivered to people who offend across the UK and Canada rely heavily on this idea of manualised treatment (Marshall, 2009). However, many researchers and practitioners have objected to this as restricted practice which hinders therapeutic responsiveness and flexibility (Addis & Krasnow, 2000; Beutler, 1999). Marshall (2009) reviewed the impact of utilising detailed manuals in the delivery of treatment to people who have committed a sexual offence and conceptualises the negative impact of detailed manuals as being twofold. Firstly, in order to evaluate the impact that this type of treatment has on future recidivism, programmes must be run over many years with high numbers of treatment completions. This, in turn, means that advances in the literature are hindered from being implemented because the manual must remain consistent. Secondly, Marshall argues that the strict adherence to a manualised approach reduces the effective implementation of the core principles of treatment in forensic settings (Andrews & Bonta, 2003; Gendreau et al., 1999). In basic terms, group dynamics are arguably negatively affected by an over-manualised approach because it reduces opportunities to be responsive to the various stages of development or socialisation that the group may be in at any one time. Indeed, some of the criticisms of the RNR model, and its influence on group interventions, are that it does not provide therapists with the sufficient tools needed to engage and work with people who have offended, pays limited attention to the therapeutic alliance, and limits attention to wider (non-criminogenic) needs, such as personal distress and low self-esteem (Ward et al., 2007).

Group Formation and Group Socialisation

Social psychological theories of group formation and socialisation offer helpful approaches to considering some of the challenges alluded to above. Tuckman (1965) originally proposed a four-stage model to conceptualise distinct phases that groups go through in order to reach maximum effectiveness. Tuckman added a fifth stage when revising the model (Tuckman & Jensen, 1977). The phases are as follows:

- *Forming* – group members are becoming acquainted with one another and learning more about the task at hand. This is a low conflict and low productivity stage where members are avoidant of controversy as they determine group organisation, structure, and roles. Group rules are likely to be established in this stage.
- *Storming* – characterised by more conflictual engagement. Group members may begin to question the task, how it is approached, and the roles and responsibilities of each group member. Tuckman (1965) suggests that group members can become hostile towards one another and towards a therapist or trainer as a means of expressing their individuality and opposing the formation of the group structure. It is proposed that, in this stage, members are more likely to have an emotional response to the work, particularly if goals are related to self-understanding and self-change (Bonebright, 2010). In the storming stage, there is likely to be significant resistance within interpersonal group relations.
- *Norming* – group development is mostly characterised by cohesion. Roles and norms have been established, and group members understand and appreciate each other. Tuckman (1965) proposed that, in this phase

of group formation, the group becomes an entity which group members seek to maintain and perpetuate. Neuman and Wright (1999) suggest that this stage is concerned with developing shared mental models, and establishing the most effective ways to work together.
- *Performing* – the group has developed functional role relatedness, and group members are sufficiently interconnected in a way that enables them to work both independently and collectively to complete tasks. Group members have reached adaptability and flexibility, and the structure of the group supports goal achievement. This stage is characterised by both harmony and productivity.
- *Adjourning* – as a later addition to Tuckman's (1965) original model, this stage represents closure, where the task is completed, and the group disengages or disbands. Van Vugt and Hart (2004) suggest that this stage can also occur if group members lose interest or motivation and the group dissolves.

Another social psychological concept that is of relevance when thinking about the role of groups within forensic practice is group socialisation. Moreland and Levine (1982, 1984) and Levine and Moreland (1994) moved beyond conceptualisations of group development to considering *how* groups interact. They subsequently proposed a model of group socialisation which argues that the interrelationship between the group and group members is most critical. The model assumes that group members continually evaluate one another in order to decide how rewarding the group interrelationships are, and that both parties are agents of influence (i.e. individuals and the group itself). This, in turn, affects the commitment levels of the group. Moreland and Levine suggest that such commitment levels can fluctuate over time, and, if change is significant enough, the group can reach the decision criterion, which may result in role transitioning within the group. They propose that commitment plays an important role in group effectiveness. Indeed, they argue that if an individual feels a strong commitment to a group, they are more likely to accept its core values, feel positively towards it, and increase work rate in order to attain group goals. Levine and Moreland (1994) propose that there are five phases of group membership:

- *Investigation* – An individual is a prospective member of the group, and may engage in reconnaissance where they seek potentially suitable groups that align with their goals. The group may engage in recruitment where it actively seeks out potential members. If commitment levels of both the individual and the group reach a satisfactory standard, and the entrance criterion is met, the transition of entry occurs, and the individual becomes a new group member.
- *Socialisation* – The individual group member and the group itself attempt to accommodate one another. Levine and Moreland (1994) suggest that it is during this phase that each party attempts to change the other.

Again, if an acceptable level of commitment is reached, and the acceptance criterion is met, the group accepts the group member.
- *Maintenance* – The individual has now been assimilated into the group, and the group and the individual engage in role negotiation. During this phase, the group works to assign the individual a particular role where their skills and contributions are optimised. Conversely, the individual also seeks to find the role that best fits their goals or personal needs. If this is successful, then commitment of both the group and the individual remains high. However, if this phase is unsuccessful as a result of the group member not being content with his/her role, for example, then the divergence criterion is reached, and the group member can become ostracised or marginalised. This in turn leads to the re-socialisation phase beginning.
- *Re-socialisation* – This phase is essentially one of re-negotiation where the group attempts to restore the group member's role within the group, and the individual member attempts to restore the group's contribution to his/her personal goals and needs. If this phase is successful, then commitment levels increase, and convergence may occur. However, if unsuccessful, commitment levels decrease, the exit criterion is reached, and the individual exits the group.
- *Remembrance* – The relationship between the group and the group member ends, and the group recalls the contributions made by their now former member. The group member also reminisces about their membership in the group. The commitment levels on both sides are low but there may be continued obligation and loyalty.

So, what implications do these theories have for forensic practice? The theory of group socialisation offers a working model of how group members' continual evaluation of one another, and subsequent levels of commitment, could directly inform when group members may be at risk of dropping out of the group, particularly during the socialisation and maintenance stages. In fact, an increased understanding on the part of facilitators about the shifting of commitment levels across the group socialisation stage, and the psychological stress of the storming stage, would be, in our view, responsivity enabling. However, this approach would need to involve increasing group facilitators' scope to respond to group dynamics more flexibly, which, in turn, would entail a reduction in over-manualised approaches.

Furthermore, Tuckman (1965) and Tuckman and Jensen's (1977) model of group formation may well offer a different lens through which to explore and improve attrition rates. While it is common for offending behaviour groups to begin with considering 'group rules', in which they establish acceptable forms of behaviour, this generally happens very early on in the group when they are still in the more harmonious 'forming' stage. The group then continues their interactions while being guided by trained facilitators and a facilitation manual. However, the 'storming' stage of the group is likely to cause considerable disharmony- something that the pre-agreed 'group rules' can be ill-equipped to deal with. Facilitators' task of managing this successfully is likely to be influenced by the level of training and experience of group facilitation that they have received. The current climate of delivering large scale economically viable group interventions, particularly in prisons, has meant that increasingly paraprofessionals have been employed as group facilitators (Tyler et al., 2019). Consequentially, qualified psychologists have taken up more supervisory and overseeing roles. Gannon et al.'s meta-analysis examining staff and programme variables as predictors of treatment effectiveness (Gannon et al., 2019) found that overall treatment effectiveness was improved when programmes received consistent hands-on input from a qualified registered psychologist, and staff who facilitated the programmes were provided with clinical supervision. The authors prompted consideration of staffing and programme implementation variables to achieve optimal reductions in recidivism.

Systems can also impact the prevention or breakdown of group formation. Often, serving prisoners are moved around the system, particularly if their behaviour is deemed problematic or disruptive to the regime. The availability of groups at particular points in someone's sentence can also affect whether they are in the right place motivationally to engage, and, if so, whether a programme is even available at that time. Furthermore, the higher incidence of mental ill health, substance misuse, and cognitive impairment among this population compared to that of the general population could also have an impact (Tyler et al., 2019). These factors could directly influence the degree to which someone may be able to participate in the 'forming' and 'norming' stages, and may arguably increase the level of 'storming'. Furthermore, the consequences of individuals acting out in the 'storming' stage can impact on their progression through the system, and hence the length of their detention. Therefore, additional time and skills may be required within the staff group to enable those with complex needs to engage successfully, and, at the very least, adequate clinical supervision as recommended by Gannon et al. (2019) should be provided.

The work of Tuckman (1965), Tuckman and Jensen, (1977), Moreland and Levine (1982, 1984), and Levine and Moreland (1994) offer models through which individual need can be better accounted for, and group deliverers can better understand group tension and predict which group members may dropout. Those who come into contact with the CJS find an expectation that they will be in many different groups- their membership of which they will have little autonomous choice over. Additionally, as referred to earlier, life histories and experiences are likely to significantly impact *how* groups have been experienced, and, subsequently, how safe an individual feels when thrust into group scenarios that were not of their choosing. These are important clinical issues that need addressing so that people can be supported when navigating group dynamics, attrition rates can be reduced, and the chance of risk reduction can be maximised.

Let us now consider a vignette of 'Steve', who finds himself expected to join a substance misuse treatment group, as directed by his sentence plan, which is necessary for him to demonstrate risk reduction and, ultimately, be released from detention.

> Steve has a long history of both acquisitive and violent offending. He has a history of both drug and alcohol misuse, and has served several short sentences in prison. Steve's childhood was characterised by witnessing violence within the home, where his father regularly attacked his mother. Steve was sexually assaulted by a teacher when he was eight years old.
>
> Steve is currently serving an indeterminate public protection (IPP) sentence for aggravated burglary and assault of a police officer during arrest. He has a tariff of 18 months and is now 12 months over this tariff period. Part of his sentence plan includes completing a substance misuse treatment group. Steve is one of eight group members. During the third session, he discovers that one of the other group members was convicted of assaulting his partner and breaking his jaw. While group rules have been established between group members, Steve begins to make comments about 'wife beaters' and 'faggots' during session, and becomes very angry with group members and facilitators. Steve is asked to meet with his offender manager (OM) to discuss his behaviour in the group. He is informed that he must desist from his behaviour and complete the group, or his upcoming parole hearing is likely to not support his release. Steve's OM tells him that negative comments about sexuality will not be tolerated. Steve responds angrily and punches his OM. He is charged with actual bodily harm and receives an additional sentence of 12 months detention.

Arguably, Steve failed to make it past the 'storming' stage of the group. Established group rules are unlikely to have considered the impact of a group member's offence triggering traumatic childhood memories on another group member. In addition, while group rules often include respect for others and no discriminatory comments, they rarely explore, in any great depth, what this will mean in practice. One potential way to improve the responsivity of the group is to focus the group on the stages of group formation and discuss the 'storming stage' openly. As part of this work, the impact of each person's offence history could be addressed transparently. Steve's history of childhood victimisation is unlikely to have been addressed, and, as such, group treatment delivery has not been developed to be responsive to his individual needs. In addition to this, his response in the group can be formulated as a clear reduction in his commitment levels, which can be addressed in a variety of ways, both outside and inside the group. For example, it is likely that being instructed to stop his behaviour in order to avoid facing a negative Parole outcome will not have aided Steve in addressing his commitment levels and working through his own internal responses. This could be more effectively managed by an OM intervening with a more motivational interviewing style.

Of course, the system only has a limited number of resources to work with. However, working more effectively and responsively could reduce potential harm and save time. Additionally, Steve's individual responsivity needs are very likely to be the reason that he is over tariff in his original sentence. He now has an additional sentence to serve due to assaulting his OM. This has significant human impact for Steve, his family, and the OM, but also has a significant implication for the use of public funds.

Professional Group Decision-Making and Exit from the System

Once an individual has entered and progressed through a criminal justice system, they will face many potential exit points from it. Some of these points of exit are dependent on the type of sentence an individual is serving. For example, in the UK, if an individual has received an indeterminate sentence, they will need to successfully progress through a parole board before they can be released. Similarly, if they are detained under the Mental Health Act, there are many decision-making points that will influence an end to their detention, not least a mental health tribunal (MHT).[4] Throughout an individual's care and progression through justice systems, there are professionals involved in decision-making for that individual who are inevitably influenced by their own social identity and group membership. This section will now consider social psychological models and how they relate to staff decision-making.

Groupthink

The field of social psychology has attempted to offer explanations for many aspects of group behaviour and decision-making, including the variety of factors that can influence *how* and *why* a group reaches a consensus, and the processes that can affect the efficacy of the decision-making process more generally (Baron et al., 1992; Janis 1971, 1972, 1982; Levine & Moreland, 1994; Moreland & Levine, 1982, 1984; Myers & Lamm, 1976; Tuckman, 1965). One of the most well-known theories associated with this topic, which aims to explain poor decision-making processes and outcomes in group contexts, is Janis's theory of 'groupthink' (Janis, 1971, 1972, 1982). In brief, groupthink refers to the pursuit of, and conformity to, a consensus within a group, typically to maintain group harmony. Janis suggested that groupthink can occur when a consensus is reached too soon, as a result of premature concurrence-seeking, the presence of a strong, persuasive group leader, a high level of group cohesion, intense pressure from the outside to make a good decision, and/or when facing a 'provocative situational context' (e.g. a moral dilemma or scenario in which the group is likely to experience material losses). However, it is important to note that Janis's theory has been criticised for having poor explanatory power (e.g. it cannot explain why groupthink does not occur in *all* groups facing a problem-solving situation; Neck & Moorhead, 1992).

Additionally, groupthink can occur in situations where group members believe that it is more important to reach a unanimous decision than it is to carefully and methodically consider all courses of action (Janis, 1971, 1972, 1982). One famous, and ultimately fatal, example of this can be seen in the Space Shuttle Challenger disaster, where engineers and company executives *knew* of the faulty O-ring seals *prior* to take-off, but went ahead and authorised the launch anyway in an attempt to avoid any negative media press about delays to the space programme (Esser & Lindoerfer, 1989). Another archetypal example of groupthink can be found in President Kennedy's failed Bay of Pigs invasion, where, in the name of camaraderie, his top advisors supported and endorsed his decision to invade the south coast of Cuba, despite having serious doubts about the success of the mission (Janis, 1972, 1982). In both examples, Janis did not believe that the grievous decisions made by such highly skilled and trained experts were a result of any inherent evil, incompetence, or laziness, but rather a consequence of seeking group consensus.

Groupthink is more likely to occur when there is a strong sense of '*we*' within the group. Where this is the case, the group's emphasis is on agreement, and there is a sense of threat evident if a consensus is not reached. The group's harmony is of paramount importance, but achieved at the cost of providing objective and critical thoughts and opinions, which leads to poor decision-making, poor problem solving, and unmet goals (Janis, 1972, 1982; Rose, 2011).

There are many contexts in forensic practice where group decisions are made; decisions that can involve complex moral dilemmas and/or material losses. A number of these decisions can have a major impact on an individual's progression through the system or, conversely, on the general public when incorrect decisions are made about risk. For example, Parole Board decisions around releasing an individual from custody, or tribunal decisions related to patients detained under the Mental Health Act, could potentially be susceptible to group decision-making bias, including Janis's groupthink (Janis, 1972, 1982).

Furthermore, Janis (1982) put forward eight symptoms of groupthink (Table 7.1). The first two symptoms *(illusions of invulnerability* and *collective rationalisation)* stem from an over-confidence in the group's power. There are many examples of this within the CJS, not least of which is related to the in-built power imbalance between staff and service users (Lloyd et al., 2017). As identified in the first part of the chapter, prisoners such as Steve may face the penalty of non-progression through their sentence if they do not, or cannot, meet the recommendations made by staff.

The third and fourth symptoms of groupthink (*belief in inherent morality of the group* and *out-group stereotypes*) are characterised by the limited vision members use to view the problem (i.e. only viewing from their own viewpoint or using 'in-group' view). Arguably, the CJS is often concerned with judgements about 'right and wrong' or 'good and evil'. This can create the conditions for 'splitting' to occur in different ways. For example, when staff

Table 7.1 Symptoms of groupthink

Symptom	Description
1. Illusions of invulnerability	The groups display excessive optimism and take big risks. The members of the group feel as though they are perfect, and that anything they do will turn out to be successful.
2. Collective rationalisation	Members of the group rationalise thoughts or suggestions that challenge what the majority is thinking. They try giving reasons as to why the others do not agree and thereby go ahead with their original decisions.
3. Belief in inherent morality of the group	There is a belief that whatever the group does it will be right, as they all know the difference between right and wrong. This causes them to overlook the consequences of their decisions.
4. Out-group stereotypes	The group believes that those who disagree are opposed to the group on purpose. They stereotype them as 'weak', 'stupid', or 'evil', and incapable of making the right decisions.
5. Direct pressure on dissenters	The majority directly threaten the person who questions the decisions by telling them that they can always leave the group if they do not want to agree with the majority. Pressure is applied to get them to agree.
6. Self-censorship	People engage in self-censorship when they believe that *they* must be the one who is wrong if they are the only odd one out.
7. Illusions of unanimity	Silence from some is considered to be acceptance of the majority's decision.
8. Self-appointed mind guards	These are members of the group who take it upon themselves to discourage alternative ideas from being expressed.

polarise into holding either the victim or perpetrator parts of prisoners, those who hold the victim/vulnerability aspects risk potentially being judged as 'Care Bears' or 'True Carers' according to Tait's (2011) prison officer typology model. On the other hand, those who only attend to the perpetrator side of a prisoner with their own moral judgements of right and wrong, may potentially engage in an overuse of power (Liebling & Price, 2001). This was amply demonstrated in Zimbardo et al.'s Stanford Prison Experiment, which had to be halted prematurely due to the extreme behaviours exhibited by participants (Zimbardo et al., 1971). Specifically, 'prison guards' became increasingly sadistic, while 'prisoners' became progressively psychologically distressed and victimised. Therefore, the unchallenged belief in the morality of the group, or polarising out-group stereotypes, can result in dangerous behavioural outcomes.

The last four symptoms (*direct pressure on dissenters, self-censorship, illusions of unanimity,* and *self-appointed mind guards*) are characterised by signs of strong compliance pressure from within the group. The strength of peer pressure and group cohesion cannot be underestimated, particularly in environments where your survival may depend on your group membership, or relationship with staff and other service users. There are different ways in which dissent may be dealt with, including positional power (Steiner & Wooldredge, 2018), social conformity (Haslam & Reicher, 2012) and defensive practice in relation to risk aversion (Tuddenham, 2000). Inspections, audits, rules, and regulations can be used to bring both staff and prisoners into line. Where self-censorship and illusions of unanimity exist, there is a risk that only the extroverted and confident group members will be listened or attended to, thus losing the important input of the more introverted group members.

Groupthink is said to be more likely to occur when there is a strong, persuasive group leader, high level of group cohesion, and intense pressure from the outside to make a good decision. Within forensic practice, there are several arenas where these preconditions exist. For example, within a UK Parole Board setting, the chair of the board (commonly a member of the Judiciary) may be perceived as the 'persuasive' group leader, with other members made up of psychology, probation, and other criminal justice professionals. To mitigate the effect of strong, persuasive group leaders, it is important for leaders to develop their listening and chairing skills, while working to empower and encourage all participants to share their views. This highlights the need for people within the CJS to develop their levels of insight and responsibility (Webster et al., 2020), and recognise the importance of relationships- particularly between those of staff and prisoners (Liebling & Price, 2001).

The following vignette considers the group dynamics that may occur when staff are making decisions about the progress of a forensic mental health patient.

> Amanda has been detained in a medium secure unit (MSU) on a section 37/41. She assaulted her ex-partner's new husband after a period of stalking and harassing them both. She has a history of self-harm and suicidal ideation. Amanda was a 'looked after' child following her parent's relationship breakdown which was initiated by her father's imprisonment and mother's ill mental health. Amanda's response to staff on the unit ranges from idealisation and over-involvement to denigration and avoidance.
>
> Staff on the unit either hold the perpetrator or victim elements of Amanda. This leads to frequent disagreements between ward staff on how best to work with her, and how quickly she should progress to ground leave (i.e. escorted leave on the grounds of the hospital).
>
> During her psychology session, Amanda has disclosed that she was physically abused by her father when she was living at home, and by a male staff member at the care home who used to frequently remove her possessions and withhold treats from her as a 'punishment.'

During patient review meetings, it is usual practice for staff to discuss each patient, and review their progress prior to asking them to join to talk about any requests. Prior to Amanda joining the meeting, the Occupational Therapist (OT) and psychologist have been making the case for Amanda to have escorted leave on the grounds of the hospital, linked with a clear behavioural plan. The medical lead, ward manager, and nursing lead are opposed to this, and are refusing to allow any nursing staff to accompany Amanda due to her unpredictable behaviour and frequent refusals to take medication.

In this scenario, the different professionals have different agendas:

- The psychologist is focusing on a formulation discussion to try and understand the reasons for Amanda's presentation, and how to treat her complex trauma.
- The OT is focused on increasing Amanda's choice and volition.
- The psychiatrist is concerned with Amanda's medication compliance and the risks of prescribing sleep medication given her risk to self.
- The ward manager says that staff are anxious about Amanda acting out violently, and that nursing staff are split between a group that want to work with her and a group that are actively avoiding her.

There may be a tendency for the group decision-making about Amanda to be driven by groupthink processes. For example, the medical lead could be viewed as a strong and persuasive leader, given the hierarchical nature of medical decision-making. The group may be split into two, with each element becoming more polarised in their decision-making. It is important for them to stop and think about the objective evidence before them, and to question their decision-making. The group could consider the formulation of Amanda's presentation, being mindful of how her childhood experiences may have contributed to her behaviour as an adult. This may help the group to challenge their own feelings or responses, and consider alternative explanations to help make sense of recent behaviour and determine the best way forward. Then, it is important for the group to have a cohesive plan of action, and consistency of approach prior to inviting Amanda into the group. The group could be assisted in their decision-making by taking time to listen to one another's points of view, and attending reflective practice, or supervision, to work out how their own issues may be influencing their viewpoint.

Group Polarisation

Another important phenomenon to consider at various points of exit from the system is that of group polarisation. This refers to the shift that occurs in groups towards more extreme decisions following group discussion (Isenberg, 1986). Somewhat counterintuitively, Myers and Lamm (1976) found that judgements made *after* group discussion will be more extreme in the same direction as the average individual judgements made prior to discussion. In other words, the process of discussion in a group does not moderate people's judgements, but rather amplifies them. This has implications for a variety of group decision-making settings in the CJS. Specifically, in reference to potential points of exit from the system, group polarisation could be applicable to arenas such as multi-agency public protection arrangements (MAPPA),[5] parole board hearings, and patient review meetings in health settings.

In forensic practice, where staff decision-making can often be considered as having high stakes for failure and thereby creating high levels of anxiety (Mann et al., 2014), it is particularly important for practitioners to be mindful of phenomena such as groupthink and group polarisation. To avoid these phenomena materialising, it is important to have a process in place for checking the fundamental assumptions behind important decisions involved in validating decision-making, and evaluating the risks involved. It is important to explore objectives and alternatives, encourage the challenging of ideas, and have back-up plans. Gathering data and ideas from outside sources, and evaluating them objectively, could be achieved by an external facilitator leading reflective practice. Regular clinical supervision, reflective practice (Gannon et al., 2019; Webster et al., 2020), and trauma-informed practice (Fallot & Harris, 2009; Maguire & Taylor, 2019) are all ways in which to create a psychologically safe space to consider and question alternative ideas, and avoid closing down discussion and thinking space. The group needs to feel safe enough to learn from failure. This could include utilising external audit processes- but these must be balanced with an inquisitive and open approach.

Conclusion

In this chapter, we have considered the journey through imprisonment to release from the perspective of social psychological models and theories. While such models have tended to fall out of favour over the years, with some having critically evaluated their usefulness, they offer valuable learning points that can enrich current forensic practice. The influence of the RNR model, the increased manualisation of rehabilitation groups in forensic settings, and an over-reliance on cognitive-behavioural therapy (CBT) interventions, has had a significant impact on practitioners' freedom to be clinically responsive in rehabilitative groups. The narrow view of what constitutes 'criminogenic' need has arguably constricted practitioners' ability to address the traumatic histories of those who have found themselves under the management of the CJS, side-lining the reality that people who offend are often in pain or distress (Taylor & Hocken, 2021). It could be argued that 'criminogenic needs' are, in essence, 'symptoms' and not etiological factors (Ward & Beech, 2015). The recognition of this allows for greater freedom of intervention, which, in turn, could encourage practitioners to focus more on areas such as group formation and development, and core group dynamics that occur within these frameworks. Additionally, it opens the door to introducing different therapeutic approaches, such as compassion focussed therapy (Gilbert, 2009), schema therapy (Young et al., 2003), and trauma informed practice (Fallot & Harris, 2009). There are some signs of a positive move in this direction, such as the removal of the Sex Offender Treatment programme in UK prisons in the wake of disappointing outcome data, and the introduction of more strengths-based models ('Horizon' and 'Kaizen') that appear to be grounded in the Good Lives Model (McCartan & Prescott, 2017; Ward & Brown, 2004). While both of these programmes retain the constrictions of the RNR model, and a narrow 'criminogenic' focus, they importantly allow those who maintain their innocence to partake in the group (McCartan & Prescott, 2017).

This chapter has also explored how key social psychological theories of groupthink (Janis, 1972, 1982) and group polarisation (Isenberg, 1986) can potentially impact the effectiveness of professional decision-making at points of exit from forensic detention. Regular supervision, reflective practice, and trauma-informed practice may well offer useful safeguards against such vulnerabilities in group decision-making.

As explored in Chapter 9, stigma, isolation, and ostracism can have significant impacts on people's experiences of forensic services. As such, approaches that recognise that individuals exist in a social context, where group membership is important, and are negatively impacted by ostracism, are becoming more commonplace. This seems critical if we are to allow people who have offended to remove this stigma from their histories and contribute to their families and communities in ways that promote psychological health. A positive example of a rehabilitative community group that promotes positive connectivity is that of Circles of Support and Accountability (COSA; Wilson & Prinzo, 2001; Wilson & Pichecha, 2005), operating in North America and Europe. The Circles Model originated in the Canadian Correctional Service as a mechanism to help men convicted of sexual offences safely reintegrate back into the community. This was achieved by creating a 'circle' of support in which trained volunteers from the community worked with the individual. The goal of COSA is 'to promote successful integration of released men into the community by providing support, advocacy, and a way to be meaningfully accountable in exchange for living safely in the community' (Correctional Service of Canada, 2002, p. 7). Efficacy research for the COSA model reports positive results in reducing reoffending. COSA Canada report that studies conducted between 2005 and 2009 show that participants committed between 70 and 83 per cent fewer offences than those who did not participate in the programme (Wilson et al., 2009). Perhaps one of the main effects is achieved through improving individuals' chances of successfully changing their group membership from 'ex-offender' (a term used explicitly by the service) to that of 'citizen' community member.

Notes

1 Adverse childhood experiences include: parental separation/divorce, verbal abuse, household mental illness, physical abuse, domestic abuse, household alcohol abuse, sexual abuse, household drug abuse, household member incarcerated, emotional neglect, and physical neglect.
2 Programme 'starts' refers to the number of individuals who began a group programme. It offers no information on how many people complete, or attrite, the group.
3 An offender manager is responsible for the person with a conviction throughout the whole of their sentence. They are responsible for assessing the person's risks and needs, planning how their sentence should run, and/or deciding upon necessary interventions, such as rehabilitation groups.
4 The MHT is the 'court' which convenes at the hospital at which a patient is detained under the UK Mental Health Act, and which determines whether the grounds for detention under the Act continue to exist. The panel consists of three members: a medical member, a legal member and a lay member.
5 MAPPA refers to the process through which different agencies, including the police and HM Prison and Probation Service (HMPPS), work together to reduce the risks associated with people who have committed a sexual or violent offence and are living in the community.

References

Addis, M. E., & Krasnow, A. D. (2000). A national survey of practicing psychologists' attitudes toward psychotherapy treatment manuals. *Journal of Consulting & Clinical Psychology*, 68, 331–339.

Ambrose, D. (2005). Cognitive skills groupwork. In D. Crighton & G. Towl (eds), *Psychology in probation services* (pp. 91–103). BPS Blackwell.

Andrews, D. A., & Bonta, J. (2003). *The psychology of criminal conduct* (3rd edition). Anderson.

Andrews, D.A., & Dowden, C. (2006). Risk principle of case classification in correctional treatment. *International Journal of Offender Therapy and Comparative Criminology*, 50, 88–100.

Andrews, D. A., Zinger, I., Hoge, R. D., Bonta, J., Gendreau, P., & Cullen, F. T. (1990). Does correctional treatment work? A clinically relevant and psychologically informed meta-analysis. *Criminology*, 28(3), 369–404.

Andrews, D. A., Bonta, J., & Wormith, J. S. (2006). The recent past and near future of risk and/or need assessment. *Crime and Delinquency*, 52, 7–27.

Baron, R. S., Kerr, N., & Miller, N. (1992). *Group process, group decision, group action*. Thomson Brooks/Cole Publishing.

Beutler, L. E. (1999). Manualizing flexibility: The training of eclectic therapists. *Journal of Clinical Psychology*, 55, 399–404.

Bonebright, D. A. (2010). 40 years of storming: a historical review of Tuckman's model of small group development, *Human Resource Development International*, 13 (1), 111–120.

Bonta, J., & Andrews, D. A. (2007). Risk-need-responsivity model for offender assessment and rehabilitation. *Rehabilitation*, 6(1), 1–22.

Cann, J., Falshaw, L., Nugent, F., & Friendship, C. (2003). *Understanding What Works: Accredited cognitive skills programmes for adult men and young offenders*. Research Findings no. 226. Home Office.

Correctional Service of Canada. (2002). *Circles of support and accountability: A guide to training potential volunteers*. Correctional Service of Canada.

Duwe, G. (2017). *The use and impact of correctional programming for inmates on pre- and post-release outcomes*. National Institute of Justice.

Esser, J. K., & Lindoerfer, J. L. (1989). Groupthink and the space shuttle Challenger accident: Toward a quantitative case analysis. *Journal of Behavioural Decision Making*, 2, 167–177.

Fallot, R., & Harris, M. (2009). *Creating cultures of trauma-informed care (CCTIC): A self-assessment and planning protocol*. Community Connections.

Falshaw, L., Friendship, C., Travers, R., & Nugent, F. (2004). Searching for "what works": HM Prison Service accredited cognitive skills programmes. *The British Journal of Forensic Practice*, 6, 3–13.

Gannon, T. A., Oliver, M. E., Mallion, J. S., & James, M. (2019). Does specialized psychological treatment for offending reduce recidivism? A meta-analysis examining staff and program variables as predictors of treatment effectiveness. *Clinical Psychology Review*, 73, 1–18.

Garrett C. J. (1985). Effects of residential treatment on adjudicated delinquents: A meta-analysis. *Journal of Research in Crime and Delinquency*, 45, 287–308.

Gendreau, P., & Andrews, D. A. (1990). Tertiary prevention: What the meta-analyses of the offender treatment literature tell us about 'what works'. *Canadian Journal of Criminology*, 32(1), 173–184.

Gendreau, P., Goggin, C., & Smith, P. (1999). The forgotten issue in effective correctional treatment: Program implementation. *International Journal of Offender Therapy and Comparative Criminology*, 43, 180–187.

Gilbert, P. (2009). Introducing compassion-focused therapy. *Advances in Psychiatric Treatment*, 15(3), 199–208.

Harvey, J. (2011). Acknowledging and understanding complexity when providing therapy in prisons. *European Journal of Psychotherapy and Counselling*, 13(4), 303–315.

Haslam, S. A., & Reicher, S. D. (2012). When prisoners take over the prison: A social psychology of resistance. *Personality and Social Psychology Review*, 16(2), 154–179.

Isenberg, D. J. (1986). Group polarization: A critical review and meta-analysis. *Journal of Personality and Social Psychology*, 50(6), 1141–1151.

Izzo R. L., & Ross R. R. (1990). Meta-analysis of rehabilitation programmes for juvenile delinquents: a brief report. *Criminal Justice and Behaviour*, 17, 134–142.

Janis, I. L. (1971). Groupthink. *Psychology Today*, November, 43–84.

Janis, I. L. (1972). *Victims of groupthink: A psychological study of foreign-policy decisions and fiascoes*. Houghton-Mifflin.

Janis, I. L. (1982). *Groupthink: Psychological studies of policy decisions and fiascoes* (2nd ed.). Houghton-Mifflin.

Levine, J. M., & Moreland, R. L. (1994). Group socialization: Theory and research. *European Review of Social Psychology*, 5(1), 305–336.

Liebling, A., & Price, D. (2001). *The prison officer*. Prison Service and Waterside Press.

Lipsey, M. W. (1992). Juvenile delinquency treatment: a meta-analytical inquiry into the variability of effects. In T. D. Cook, H. Cooper, D. S. Cordray, H. Hartmann, L. V. Hedges, R. J. Light, T. A. Louis, & F. Musteller (eds), *Meta-analysis for explanation: A casebook* (pp. 83–125). Russell Sage Foundation.

Lloyd, C., Page, G. W., Liebling, A., Grace, S. E., Roberts, P., McKeganey, N., Russell, C., & Kougiali, Z. (2017). A short ride on the penal merry-go-round:

Relationships between prison officers and prisoners within UK drug recovery wings. *Prison Service Journal*, 230, 3–14.

Maguire, D., & Taylor, J. (2019) A systematic review on implementing education and training on trauma-informed care to nurses in forensic mental health settings. *Journal of Forensic Nursing*, 15(4), 242–249.

Mann, B., Matias, E., & Allen, J. (2014). Recovery in forensic services: Facing the challenge. *Advances in Psychiatric Treatment*, 20(2), 125–131.

Marshall, W. L. (2009). Manualization: A blessing or a curse? *Journal of Sexual Aggression*, 15(2), 109–120.

Marshall, W. L., Marshall, L. E., & Burton, D. L. (2013). Features of treatment delivery and group processes that maximise the effects of offender programs. In J. L. Wood & T. A. Gannon (eds), *Crime and crime reduction: The importance of group processes* (pp. 159–176). Routledge.

McCartan, K., & Prescott, D. (2017). Bring me the horizon! (and kaizen). Journal of Sexual Abuse Blog. Retrieved from www.davidprescott.net/articles/20170629SAJRT.pdf

McMurran, M., & Theodosi, E. (2007). Is treatment non-completion associated with increased reconviction over no treatment? *Psychology, Crime & Law*, 13(4), 333–343.

Messina, N. P., & Schepps, M. (2021). Opening the proverbial 'can of worms' on trauma-specific treatment in prison: The association of adverse childhood experiences to treatment outcomes. *Clinical Psychology & Psychotherapy*, 28, 1–12.

Ministry of Justice. (2015). *Accredited programmes annual bulletin 2014/15: England and Wales Ministry of Justice statistics bulletin*. Ministry of Justice.

Moreland, R. L., & Levine, J. M. (1982). Socialisation in small groups: Temporal changes in individual-group relations. In L. Berkowitz (ed.), *Advances in experimental social psychology* (pp. 137–192). Academic Press.

Moreland, R. L., & Levine, J. M. (1984). Role transitions in small groups. In V. L. Allan, & E. Van De Vliert (eds), *Role transitions: Explorations and explanations*, (pp. 181–195). Plenum.

Myers, D. G., & Lamm, H. (1976). The group polarization phenomenon. *Psychological Bulletin*, 83, 602–627.

National Offender Management Service. (2005). *Annual report for accredited programmes, 2004–2005*. Home Office.

Neck, C. P., & Moorhead, G. (1992). Jury deliberations in the trial of US v John DeLorean: A case analysis of groupthink avoidance and an enhanced framework. *Human Relations*, 45, 1077–1091.

Neuman, G. A., & Wright, J. (1999). Team effectiveness: Beyond skills and cognitive ability. *Journal of Applied Psychology*, 84(3), 376–389.

Rose, J. D. (2011). Diverse perspectives on the groupthink theory: a literary review. *Emerging Leadership Journeys*, 4(1), 37–57.

Ross, R. R., Fabiano, E. A., & Ross, R. D. (1986). *Reasoning and rehabilitation*. University of Ottawa.

Steiner, B., & Wooldredge, J. (2018). Prison officer legitimacy, their exercise of power, and inmate rule breaking. *Criminology*, 56, 750–779.

Tait, S. (2011). A typology of prison officer approaches to care. *European Journal of Criminology*, 8(6), 440–454.

Taylor, J., & Hocken, K. (2021). People hurt people: reconceptualising criminogenic need to promote trauma sensitive and compassion focussed practice. *The Journal of Forensic Practice*, 23(3), 201–212.

Towl, G., & Crighton, D. (2016). The Emperor's New Clothes. *The Psychologist*, 29(3), 188–191.

Towl, G., & Podmore, J. (2019). Still in denial? The sex offender treatment industry. Retrieved from www.thejusticegap.com/still-in-denial-the-sex-offender-treatment-industry/.

Tuckman, B. W. (1965). Developmental sequence in small groups. *Psychological Bulletin*, 63(6), 384–399.

Tuckman, B.W., & Jensen, M. A. C. (1977). Stages of small-group development revisited. *Group and Organization Studies*, 2(4), 419–427.

Tuddenham, R. (2000). Beyond defensible decision-making: towards reflexive assessment of risk and dangerousness. *Probation Journal*, 47(3), 173–183.

Tyler, N., Miles, H., Karadag, B., & Rogers, G. (2019) An updated picture of the mental health needs of male and female prisoners in the UK: Prevalence, comorbidity, and gender differences. *Social Psychiatry and Psychiatric Epidemiology*, 54(9), 1143–1152.

United Nations. (2018). Prejudice and discrimination: Barriers to social inclusion. Retrieved from www.un.org/development/desa/dspd/2018/02/prejudice-and-discrimination/.

Van Dieten, M., & MacKenna, P. (2001). *Moving on facilitator's guide*. Orbis Partners.

Van Vugt, M., & Hart, C. M. (2004). Social identity as social glue: The origins of group loyalty. *Journal of Personality and Social Psychology*, 86, 585–598.

Ward, T., & Beech, A. R. (2015). Dynamic risk factors: A theoretical dead-end? *Psychology, Crime & Law*, 21(2), 100–113.

Ward, T., & Brown, M. (2004). The good lives model and conceptual issues in offender rehabilitation. *Psychology, Crime, & Law*, 10, 243–257.

Ward, T., Melser, J., & Yates, P. M. (2007). Reconstructing the Risk–Need–Responsivity model: A theoretical elaboration and evaluation. *Aggression and violent behavior*, 12(2), 208–228.

Webster, N., Doggett, L., & Gardner, S. (2020). If you want to change the world you have to start with yourself: The impact of staff reflective practice within the offender personality disorder pathway. *Probation Journal*, 67(3), 283–296.

Williams, A. (2005). Group-work based interventions. In D. Crighton & G. Towl (eds), *Psychology in probation services* (pp. 82–90). Blackwell.

Wilson, R. J., & Pichecha, J. E. (2005). Circles of support and accountability: engaging the community in sexual offender risk management. In B. K. Schwartz (ed.), *The sex offender: Issues in assessment, treatment, and supervision of adult and juvenile populations* (pp. 13. 1–13. 21). Civic Research Institute.

Wilson, R.J., & Prinzo, M. (2001). Circles of support: a restorative justice initiative. *Journal of Psychology and Human Sexuality*, 13, 59–77.

Wilson, R. J., Cortoni, F., & McWhinnie, A. J. (2009). Circles of support & accountability: A Canadian national replication of outcome findings. *Sexual Abuse*, 21(4), 412–430.

Wong, S. C., & Gordon, A. (2013). The violence reduction programme: A treatment programme for violence-prone forensic clients. *Psychology, Crime & Law*, 19(5–6), 461–475.

Young, J. E., Klosko, J. S., & Weishaar, M. E. (2003). *Schema therapy*. Guilford Press.

Zimbardo, P. G., Haney, C., Banks, W. C., & Jaffe, D. (1971). *The Stanford prison experiment*. Zimbardo Incorporated.

8 Coercion and Social Influence

Vyv Huddy and Timothy A. Carey

Coercion, or using force or threats to "persuade" people to engage in behaviours they would otherwise not undertake, has been an enduring feature of human interactions. Slavery has existed for millennia and continues to this day in a multitude of forms (United Kingdom Home Office, 2019). Other forms of coercive practices occur in a wide variety of institutions: healthcare settings, schools, universities, government and commercial industry, thought they are perhaps most noticeable in secure criminal justice and mental health settings. In 2016 it was revealed that institutionalised teenagers in Australia's Northern Territory had been hooded, restrained, exposed to teargas, and placed in solitary confinement. Prime Minister Malcolm Turnbull ordered a royal commission into these events, which recommended the closure of the Don Dale Youth Detention centre where the abuse took place as well as changes in the law (Royal Commission to the North Territory Government, 2017). Commentators compared the footage of the abuse of these children to events at Guantánamo Bay or the Abu Ghraib (Davidson, 2016), which highlights how institutional abuse can become part of the mainstream care of children. Coercion is also a major societal problem – in the UK, coercive behaviour is now a crime under the Serious Crime Act (2015) carrying a sentence of up to five years imprisonment. A vital question for policy makers, criminal justice professionals, clinicians, and society more widely is what situations and circumstances allows this coercive abuse to arise and continue?

This chapter opens with a definition of coercion, which highlights two key features. The first is that coercion is a subjective experience and the second is that it is an aversive state that people sometimes resist and struggle against. We explain that established perspectives in criminology or behaviourism fail to acknowledge how attempts to control people's behaviour are frequently met with non-compliance or resistance: people frequently *countercontrol* those who attempt to control them. We explain that countercontrol is the result of attempts to control the behaviour of others without taking their goals and purposes into account. This then introduces the main focus of the chapter, which is to consider coercion from the perspective of a theory that defines behaviour by purpose. In this view the purpose of behaviour is to

DOI: 10.4324/9781315560243-8

achieve or maintain perceptual states in desired or preferred conditions. We argue this definition of behaviour provides a clear means of understanding why social interactions can be experienced as either abusive or coercive. We will explain several implications of this perspective for forensic practice and how it might offer an opportunity for criminal justice experienced as humane, while also being effective and efficient.

Defining Coercion

The *Oxford English Dictionary* defines coercion as "persuade (an unwilling person) to do something by using force or threats". The reference to "unwilling" indicates that the perspective of the unwilling person is key to the definition of what is coercive. In the mental health field, the notion of "perceived coercion" (Newton-Howes & Mullen, 2011) has been introduced to distinguish "objective, external acts and internal, subjective attitudes" (p. 465). The definition of coercion used in the psychiatric literature acknowledges that coercion is "best thought of as an internal subjective state, commonly referred to as 'perceived coercion'" (ibid., p. 465). As is plain from the definition above, a key aspect of the experience of coercion is that it is a negative, unwanted, situation from the perspective of the coercee. Indeed, a systematic review of the literature on coercion in psychiatric settings identified a series of negative themes such as feeling dehumanised or unheard (ibid.).

Coercion in Prisons

In the United States the number of legally mandated interventions has increased in the past twenty years (Parhar et al. 2008). This increase may reflect the opinion that mandated intervention increases compliance and completion of treatment (Coviello et al. 2013). However, a meta-analysis on this topic (Parhar et al., 2008) found that mandated interventions were less effective, particularly when carried out in custodial settings.

Prison staff report that they work in challenging environments where assault and injury are commonplace with over 5000 assaults recorded in UK prisons in 2016 (Ministry of Justice, 2017). Seclusion and restraint are used for custodial management purposes to minimise damage to estate property as well as injury to staff and other prisoners. Restraints vary across countries but can be highly restrictive. For example, in many countries steel handcuffs, leg irons, and chair restraints are used (United Nations, 2013). These methods are applied when a prisoner is disruptive, assaultive, or otherwise posing a danger. There have been a series of high-profile incidents in the UK where people have died during police custody as a result of inappropriate restraints (Angiolini, 2017). Physical coercion via seclusion or restraint can be particularly problematic when applied by staff with little training (Champion, 2007). However, physical coercion can occur unofficially and, apparently, spontaneously. A quote from a study carried out in the USA by Marquart (1986, p. 351) strikingly illustrates this point:

> I [hall officer] had a hard time in the North Dining Hall with an inmate who budged in line to eat with his friend. Man, we had a huge argument right there in the food line after I told him to "Get to the back of the line." I finally got him out [of the dining hall] and put him on the wall. I told my supervisor about the guy right away. Then the inmate yelled "Yea, you can go ahead and lock me up [solitary] or beat me if that's how you get your kicks." Me and the supervisor brought the guy into the Major's office. Once in the office, this idiot [inmate] threw his chewing gum in a garbage can and tried to look tough. One officer jumped up and slapped him across the face and I tackled him. A third officer joined us and we punched and kicked the shit out of him. I picked him up and pulled his head back by the hair while one officer pulled out his knife and said, "You know, I ought to just go ahead and cut your lousy head off."

This quote illustrates how unofficial coercive measures might be applied in a prison setting. Marquart's research also highlighted how a proportion of guards for the purpose of group cohesion, status, and promotions applied these practices. Moreover, there was evidence from this research that those managing the officers rewarded them for their coercive and punitive behaviour. This could result in promotion or selection for easier roles within the institution.

A Criminological Perspective on Coercion

Colvin, Cullen, and Vander Ven (2002) argue that coercion occurs when individuals behave in a manner desired by others via the threat of force or other impersonal or social pressures. They describe how factors are generated from peers, family, criminal justice, social services, or the broader society. They argue that coercion can vary across two dimensions: the strength of the coercive force; and the consistency of that force. When coercion is erratic, it lowers expectations of being able to control outcomes. Effort is typically required to monitor and sanction actions (Colvin, 2000). For this reason, Colvin acknowledges that coercion is typically applied inconsistently as a result.

Colvin et al. (2002) also differentiate interpersonal from impersonal coercion. The latter relates to large structural arrangements in society that produce an indirect experience of coercion. The former involves the use of force or intimidation aimed at creating compliance at an interpersonal level. This may involve the actual or threatened action that would cause pain or the withdrawal of resources:

> Societies that allow high levels of structural unemployment and promote an ethic of 'dog eat dog' competitive individualism create the basis for multiplying the coercive networks that spin more and more individuals into the vicious cycles that create high rates of crime.
>
> (Colvin et al., 2002, p. 32)

In the past decade the UK government Home Office has implemented a "hostile environment" policy via the 2014 Immigration Act (Independent Chief Inspector of Borders and Immigration, 2016). This policy encouraged multiple agency working, including housing, vehicle regulation, insurance, financial, and healthcare services, to challenge the living conditions for those residing unlawfully in the UK. The aim is to encourage these individuals to reduce their effort to settle in the UK and return to their country of birth. While it is unclear whether this policy has had the desired impact on illegal immigration targets, there is an emerging consensus that a side effect of the policy is greater marginalisation of a population who are vulnerable to exploitation (Liberty, 2018). The policy also impacted on people from the Windrush generation who had been resident in the UK for decades and faced deportation (Taylor, 2018).

In contrast to this, Colvin et al. (2002) describes how structural coercion may also manifest in humanitarian ideologies, drawing on previous ideas of Foucault (1977). This perspective draws on historic examples of reformers apparently benevolent intentions being ultimately driven by class or religious interests. Reformers may expect appropriate behaviour in return for resources. We will later expand on the implications of assuming what is in the interests of those who seek help and how such assumptions can yield experiences of coercion.

A Behaviourist Perspective on Coercion

Coercion theory (Patterson, 1982, p. 85) took a traditional behaviourist position in understanding behaviour as "stimulus that elicits it and the consequences that maintain it. Events are conceptualised as either strengthening or weakening the stimulus response (S-R) bond". The theory further suggests that cycles of interaction exist between children and caregivers which begin when a child reacts negatively to a caregiver's request or direction. This evokes anger in the caregiver, which escalates for a time but, eventually, the parent withdraws or "gives in" to the child's purported coercion. Thus a "reinforcement trap" occurs where the parents' attempts to elicit the behaviour they prefer creates a punishment for the parent (Dishion & Patterson, 2006). The parents' response to this punishment is to allow the child to behave as they wish creating a short-term reinforcement for the parents. However, this presents longer-term reinforcement costs to the parents in the child's on-going behaviour, which they find aversive. This "coercion" model has been applied beyond parent-child interactions and informed parenting interventions for conduct problems in early and middle childhood (Eyberg, Nelson, & Boggs, 2008). The relationships between staff and adult prisoners in custodial settings could also be characterised in terms of short-term reinforcement and long-term punishments. A short-term reinforcement may be to forcibly place a prisoner behind their cell door but in the long term it might result in continued management problems.

Skinner outlined similar circumstances to those described above in his classic text *Science and Human Behavior* (1953). In this text he suggested techniques and methods to enable people to attempt to control the behaviour of others. Despite Skinner's interest in developing a method of behavioural control, he was well aware of the problems this could create. Skinner (1953) argued that, controlling the behaviour of another is likely to be experienced aversively by them and, because "of the aversive consequences of being controlled, the individual who undertakes to control other people is likely to be counter-controlled by all of them ... the opposition to control is likely to be directed to the most objectionable forms-the use of force and conspicuous instances of exploitation, undue influence" (p. 321).

Carey and Bourbon (2004) define countercontrol as "action taken by a controllee to systematically produce behavioural effects in a controller" (p. 4). Thus, the controllee and the controller are both considered to be controlling entities. However, they may not be able to control what they want to, as Marken and Carey (2015) state, "it is impossible for people to be in control and be controlled at the same time".

Control

There is a fundamental property of living things that is all encompassing and actually gives rise to the occurrence of countercontrol itself – this is the phenomenon of control itself. Control has been described as life defining: "Life is control – an uninterrupted process of specifying, creating, and maintaining – a process in which all that is not essential is free to change, preventing change in what is essential" (Bourbon, 1995, p. 151). Control can also be considered to be: "A real, objective phenomenon that involves the production of consistent results under varying environmental conditions" (Marken, 1988, p. 196).

Perceptual Control Theory (PCT) describes control as a process of acting to bring perceptions of the world to match a preferred or reference state for these perceptions. Powers (1998) stated "perception is therefore the current status of whatever it is we are trying to control" (p. 7). An automobile driver might, for example, be aware of a white line on the right of their windscreen. The position of the white line relative to the windscreen as viewed by the driver can be altered or disturbed by factors such as gusts of wind, bumps in the road, or the passing of nearby vehicles. Any of these may perturb the distance between the white line and the edge of the windscreen from the driver's perspective. However, the driver's actions counter these disturbances to ensure there is a constant distance between their windscreen and the white line – keeping it to the right side. This is how this relationship *should* look to the driver – it is their reference state for their perception of the relationship between the white line and their window. Differences between the reference value and the perception are "errors" in PCT; they are deviations between what is intended and what is perceived. Whenever there

170　*Vyv Huddy & Timothy A. Carey*

is error, this drives changes to actions such as movements on the steering wheel. Thus, the basis of action is error, or difference, from a reference state. The organisation of perception, reference, error signal, and disturbance are illustrated in Figure 8.1.

In PCT intention is "what is to be brought into experience or is to continue to be experienced" (Powers, 1998, p. 8). The effect of action is to change perception. For example, the movement of the steering wheel corrects a small deviation of the position of the window relative to the white line. Thus, the relationship between the white line and the window is, from the perspective of the driver, under control. There is a closed loop of perception to comparison to action and back to perception again. For the entire time that a living control system lives, it is controlling perceptions.

In PCT, actions are produced as a combination of the difference between perceptions and reference values along with disturbances currently occurring in the environment. As noted by Marken (1991), a living "control system controls its own inputs, it is not controlled by those inputs" (p. 4). This view of

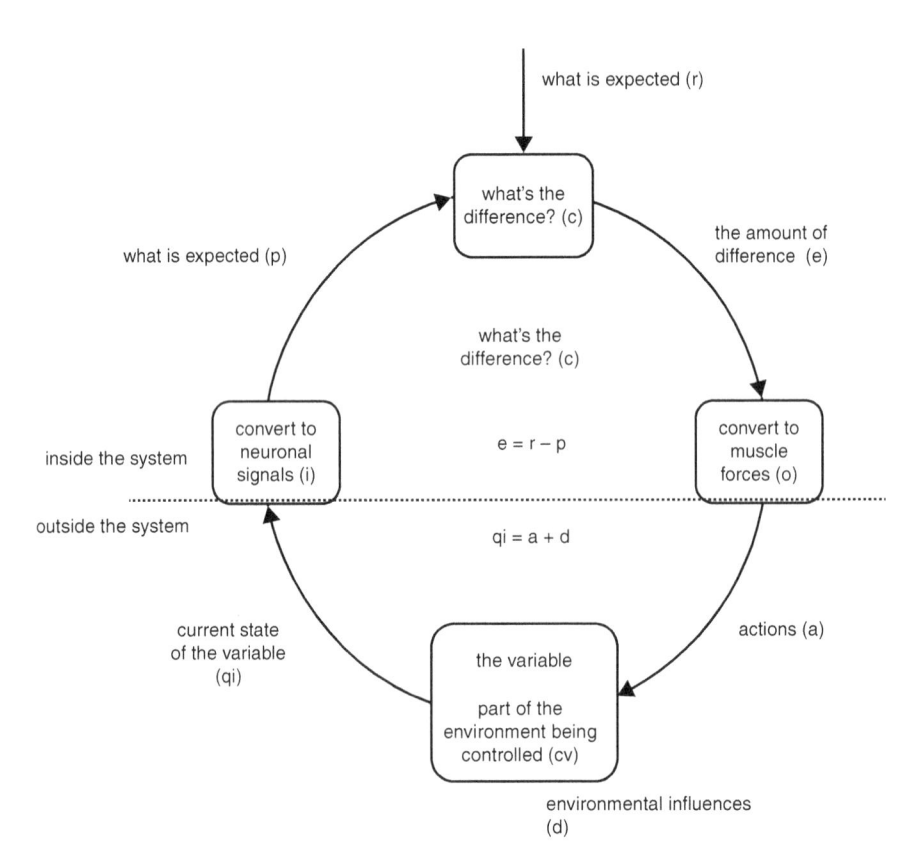

Figure 8.1 The basic feedback loop

behaviour differs fundamentally from the behaviourist view described earlier. The behaviourist's position suggests that the environment acts on all living things, including human beings, to elicit behaviour. An air puff *causes* the eyelid to close momentarily. This view is prominent in more a-theoretical assumptions in psychological science. For example, there is a widespread view that abuse or neglect causes people to have complaints with their mental health (Read et al., 2005) or a propensity to violence (Braga, Cunha & Maia, 2018). As already noted, there is a view that an experience of being coerced makes it more likely that someone will behave in a coercive manner (Colvin et al., 2002).

Considering Social Interaction from a PCT Perspective

The difficulty with understanding that behaviour is caused by events in the environment is that this neglects the controlling nature of behaviour – an abuser or coercer is a controlling entity. When the phenomenon of control and its implications for social interactions are understood it can be appreciated that, in a coercive relationship, the coercee is also controlling. The coercer and the coercee are both closed loop control systems where the coercer is in the environment of the person being coerced and the coercee is in the environment of the coercer. Thus, PCT considers any interaction between two people as the linking of two control systems through their shared environment (Bourborn, 1995). In this situation, there are at least two perceptions under control, two references defining what is controlled, and two people acting to effect their perceptions. Furthermore, the actions of each person not only effect their own perceptions, but they also affect the perceptions of the other person. It may be, however, that the actions of one person have no effect on the perceptions being controlled by the other person. A full illustration of how an interaction between two people can work is depicted in Figure 8.2.

An advantage of the PCT framework is that it is an account of all behaviour, even behaviour that appears, superficially, to be trivial. To further clarify how PCT conceives of behaviour as control it is instructive to begin with a basic example. Powers (1973) describes a paradigmatic instance of control as "the tracking experiment where a participant manipulates a control lever to cause a cursor – say a moving spot of light – to track a moving target" (p. 44). According to Powers, "this is clearly a control task. [The participant] is trying to keep the spot and the moving target in a particular relationship, namely, on the target" (ibid.). If the dot moves from the target, the participant acts to reduce the difference to zero. This task records moment-by-moment intra-individual correlations obtained between the controlled variable (e.g. location of the cursor relative to a target), the output (e.g. movements of the hand), and the disturbance (e.g. the disturbance to the location of the target). In particular, it is found that the disturbance and the behavioural output of the individual have a very high negative correlation (in excess of –0.99) because the individual's action exactly counteracts the disturbance in order to maintain control. This task allows the control of perception to be observed in an experimental situation.

172 Vyv Huddy & Timothy A. Carey

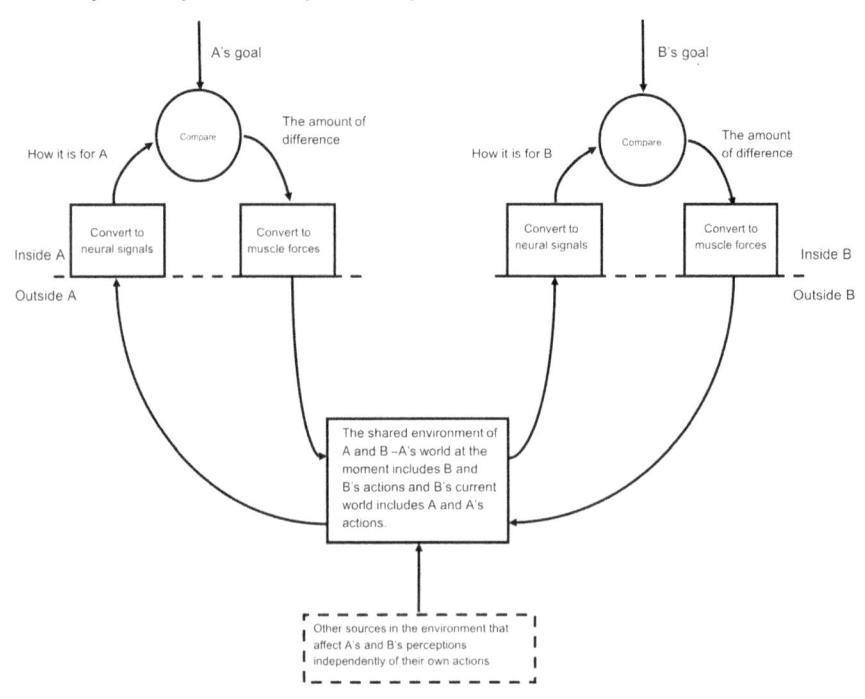

Figure 8.2 Interactions between closed casual loops

Bourbon (1990) described experiments where two individuals perform a tracking task in parallel. A task such as this allows the interaction of two people to be observed while they co-operate or interfere with each other's controlling. The study is set up so that each participant controls a separate cursor to target relationship. The crucial feature is the linkage between the participant's actions and their respective cursors. One participant's action is linked not only to their own cursor but also to that of the other participant in the experiment. Thus, an effect of their actions is a change to the position of the other participant's cursor. The striking finding is that this has no effect on the other participant's ability to control the relationship between their cursor and target. They continue to act in order to counteract the disturbances to the position of their cursor in the same way they would to any other disturbance to it. The *origin* of the disturbance is of no importance as long as it can be countered, and the preferred perceptual state maintained. This may seem an esoteric point, but it is of fundamental importance for understanding human interactions. It highlights that another person's actions can disturb the perceptions of the person they are interacting with, but this is not *necessarily* a problem for the person. In fact, they may not even notice it. It does, however, become a problem when the actions of the other person limit their ability to control what is important to them, which is a point we will return to shortly.

These experiments have been tested by the construction of *functional* or *generative* models, meaning they reproduce or simulate the behaviour observed in experiments when pairs of participants interact. Bourbon (1990) found that a computer model that implemented two interacting control loops, as described above, was a very precise fit to the experimental data where two people interacted on a task of this type, the correlation between the model and the first individual was high (r = 0.995) and similar levels of fit were observed for the second person (r = 0.996).

Implications of PCT Models for Understanding Social Interactions

The experimental demonstration, together with the simulations, has important implications for understanding social interactions. As Bourbon (1990) points out, one consideration is that this simulation demonstrates how co-ordination between two people does not require goals to be "shared" (e.g. Butterfill, 2012). Apparent co-ordination can be achieved by completely separate control systems with unique perceptions of their current status, separate goals or desires, and separate motor systems for implementing actions. The environmental linkages between the two individuals that are implemented by the computer are the only components of the interacting systems that are shared. Nonetheless, each individual still seeks to reduce the difference between *his or her own* perceptions and goals.

The discovery that a simple organisation involving two linked control systems is capable of producing apparently complex behaviour has profound implications for understanding the examples of coercive behaviour described earlier. One implication is that environmental linkages can be so organised to allow two individuals to act in ways that "disturb" each other's perceptions. When these are easily countered so that the variable is quickly restored to the reference value, disturbances are not bothersome or problematic. Indeed, the participant may not even notice the other person's presence. However, exactly the same underlying principles can also be used to understand situations where one person's controlling produces disturbances to perceptual variables that entails another person experiencing a great deal of effort and discomfort in cancelling.

The foregoing discussion explains how a continuous process of perception, comparison, and action operating in a closed loop allows goals to be achieved and preferred conditions maintained. PCT also provides an explanation of situations where control is not achieved or when important perceptual variables are, despite the efforts of the controllee, consistently not at their desired reference points. Controllees under these circumstances are, therefore, in a chronic state of error with distress or discomfort a consequence of this. The structure of the loop in Figure 8.1 highlights two possible ways environmental conditions can disrupt a person's ability to control perceptual input. Firstly, it may be that reliable information on the state of a variable is not available, such as when searching for the location of a keyhole in pitch darkness. Secondly, an environment may be configured so that output – muscular movements, speech, or writing – have no effect on perceptual variables the controllee wishes to influence. A prisoner held in a restrained position cannot physically sit up because the actions of the restraining officers are an *insuperable disturbance* to their actions. The prisoner's muscles have reached their maximum output and the disturbance created by the officers' actions is still greater than this, so they cannot move to their preferred position.

Distinguishing Coercion and Force

The concept of insuperable disturbances allows the distinction between experiences of perceived abuse and perceived coercion to be understood. Force refers to cases where an individual experiences a perceptual state, but because of an insuperable disturbance, even their *best efforts* are unable to cancel the effects of this disturbance. In contrast, coercion refers to cases where an unwilling coercee has to act in ways determined by the coercer. The examples at the outset of this chapter are perhaps more accurately considered to be examples of abuse rather than coercion. When young people at Don Dale Youth Detention Centre were hooded and restrained there were unable to act to change their perceptions. They didn't place the hoods over their heads; the hoods were fixed over their heads, as were the chair restraints. Introducing conditions where individuals are presented with insuperable disturbances denies their purpose and is the reason why people describe these experiences as de-humanising.

As noted above a coercer or abuser are themselves controlling entities, who are acting on their environment to correct disturbances to preferred states. One reason restraint of prisoners is justified is that it ensures the safety of other people or prevents damage to property. Newton-Howes and Mullen (2011) suggest that patients may experience treatment as coercive as a "side effect of the actions taken by the health care professionals" (p. 465). Invoking the notion of "side effects" draws attention to a distinction between intended and unintended effects of behaviour that follows from the PCT model of how behaviour works. Once again, the simple example of the dyadic tracking task is helpful in defining what is intentional or otherwise. A participant's actions while tracking a target may be disturbing the cursor of the other participant but this is entirely unintended – from their perspective the consequences of their actions move only their own cursor, the concurrent effects on the cursor of the other participant are unintentional or unnoticed (or both). In the same way, perceived coercion or abuse can be an unintended consequence from the perspective of a custody officer, law enforcement professional, or clinician, who are focusing solely on pursuit of their goal or goals. From their perspective, they wish to see the prisoner or patient complete their sentence plan, be safe from harm, or prevent harm to others. This, however, contradicts some other goals the prisoner may have and be experienced as coercive or abusive.

It might be argued that some degree of perceived coercion or abuse may be inevitable in a system where multiple needs must be balanced. The official may justify their actions as in the "best interests" of the prisoner or patient based on legal frameworks such as, for example, the UK Mental Capacity Act (2005). This legal provision requires the perspective of the patient or prisoner to be considered from knowledge of the person's past preferences, values, or those of significant others who can offer a substitute perspective for that of the person. These latter individuals may be appointed by court or by someone who is "most engaged" in the person's welfare. In such circumstances, the possibility that a custodial officer's decisions or actions are disturbances to the perceptions the patient or prisoner is controlling becomes more likely.

The Origins of Perceived Coercion

The scenario where someone is facing insuperable disturbance occurs when the person is acting with maximum effort to counteract the effects of the other's actions but is ultimately unsuccessful. In coercive relationships the coercee may describe himself or herself as doing something "against their will" but is not acting to counter the coercer's actions directly, at least superficially. PCT conceives of all behaviour as being purposeful so the notion of acting "against one's will" or purpose is not possible. PCT can provide an explanation for the apparent contradiction of how a prisoner completes a complex sentence plan, which incorporates extensive structured therapy, and yet they still express that they engaged in this "against their will". Understanding why people describe a certain act as "against their will" requires a further principle of PCT to be introduced, which is the principle of conflict. So far the operation of a single control system operating in isolation has been explained. However, human beings clearly pursue multitudes of goals simultaneously. In the PCT, model this multiplicity of goals is organised into a hierarchy, with more important aspects of living at the top, such as a perception of oneself as a good father, and more concrete strivings, such as taking one's daughter to football practice, situated lower down the hierarchy. Indeed, as these examples make clear, higher-level goals specify the goals lower down, so being a good father is *achieved by* setting a goal of taking one's child to sports practice.

Conflict arises when higher-level goals specify contradictory states for lower-level systems to achieve. In the example of the prisoner pursuing the sentence plan, he may be doing so to maintain a sense of himself as a good father. However, being a good father may also entail making sufficient money for some people. The prisoner may view training activities for returning to employment as critically important. In some situations, however, the scheduling for the training activities may conflict with engaging in therapeutic sessions for anger management. In this case the prisoner may feel that they are acting "against their will" and experience a perception of coercion, as they would rather spend time on employment training.

The notion that it is possible to act "against one's will" is addressed in PCT by the proposal of a multiplicity of purposes that can conflict with each other. Completion of the sentence plan requires a series of goals to be achieved: attending the sessions; reading educational materials presented by the therapists; and describing their understanding of this material when prompted to do so. These goals require just the same cycle of perception, comparison, and action as described previously. Thus, what they are doing against their "will" is also a demonstration of precise control, but control that disturbs other perceptions from the state the prisoner desires. Thus, the prisoner is not acting "against their will", one of their "wills" is acting against another of their "wills" in the multiplicity of wills that makes up one part of their neural network. The one that is most salient to them is the one that is not as they would currently like it to be. However, the need to leave prison and return to family might be more important than the need to return to work. This scenario, however, results in important unmet needs from the perspective of the prisoner and a sense that this wasn't their choice.

The Inevitability of Separate Perspectives

The inevitability that people have unique separate perspectives on decisions or choices is fundamental to a PCT understanding of human interactions. In the example described above, rehabilitation from the prisoner's perspective may be to spend all their time in detention training for return to work. However, the offender manager may take the view that completing the course of anger management sessions is the highest priority. In a healthcare setting, it has been argued that the patient experience is paramount and the foundation of effective treatment (Carey, 2016). This has been formalised into a paradigm for health services that positions the patient perspective as the key definer of treatment goals and resources (Carey, 2018).

Professionals working in prisons or forensic mental health settings may argue that the perspective of law enforcement officers, probation staff, justice experts or, indeed, the victims of crime also have a valid perspective on what rehabilitation should entail. The challenge reconciling multiple perspectives is that no matter how worthy or justified an action might be from the perspective of someone else, if it disturbs an important perceptual variable from the state that the client wants it to be in, then it will be countered by their actions. If a sentence plan to attend structured therapy disturbs how the client spends their time, then they will act to make their time spent more as they would like it to be, by not attending sessions or making their attendance as cursory as possible. If the official or clinician wishes to change the behaviour of the client towards, for example, less offending, then it is essential this is considered from the perspective of the client or patient.

How Might the PCT Perspective Improve Forensic Practice, which Minimises Coercion?

The PCT conceptualisation of behaviour as control of perceptual input means that problematic offending behaviour – for example violence, fire setting, sexual harmful acts, or theft – occurs to *counter disturbances to perceptual states* important to clients. In the same way instances of harmful coercion or abuse on the part of staff are also attempts to counter disturbances to perceptual variables important to staff. We therefore argue that an overarching principle of forensic practice is to consider the *perceptual states* that individuals in a custodial setting may be controlling rather than focusing on the problematic behaviours that are observed by others. Furthermore, this should be considered from both the perspective of the client but also the practitioners tasked with limiting the impact of their potentially harmful behaviour on others. A thorough and detailed appreciation of the perspective of the client framed in terms of perceptual variables they might be controlling ensures that practitioners can tailor interventions and management plans to minimise perceived coercion on the part of the client.

The focus on perceptual states may, on the surface, bear similarities to several other theories of offending that conceptualise the problematic behaviour as means to meet underlying needs. For example, this is evident in the applications of attachment theory (Rogers & Budd, 2016) or behaviourism (Patterson, 1982) to understand offending behaviour. A key commonality of these alternative perspectives is the assumption that the environment surrounding the individual has *caused* problematic behaviour, either as a distal or a proximal factor. As has been made clear, PCT does not consider behaviour to be caused, but is instead an outward component of the control of a perceptual variable. On these terms, knowledge of what perceptual variables the individual is trying to achieve or maintain in a particular reference state is of great value to make sense of problematic behaviour.

One way to find out about perceptual variables under control is via conversation or dialogue, allowing sufficient space and time for the client to express what is important to them. While this may run counter to some of the values of correctional establishments it would, nonetheless, be central to safe, effective, and efficient practice. In the example of the prisoner who had anger management stipulated in a sentence plan, a straightforward dialogue would have helped ascertain that he wished to increase his income. In some cases, however, learning difficulties, high levels of arousal, tension, or hostility to staff as a result of historically perceived coercion might be obstacles to dialogue. A practical, non-verbal, approach to developing an understanding of behaviour based on PCT is called the Test for the Controlled Variable (TCV) (Runkel, 1990; Carey, 2012). The TCV is a structured method for guessing what perceptual states a person might be controlling. It contrasts with the traditional behavioural functional analysis technique, which attempts to line up observations of antecedents or triggers with specific behaviours and, finally, the consequences or reinforcements of the behaviours.

The first step of the TCV is to hypothesise what variable a person is controlling by identifying the perceptual states an individual might wish to be under control. Taking a typical example of violence towards staff in a prison setting. A hypothesis could be that this behaviour maintains a state of being at a distance from other prisoners during association time. The next step is to check this out by implementing a change in environment or routine, such as offering activities during association time so that the prisoner is not on the wing. The effects of this change are then noted. If there is no change in the behaviour, then the practitioner forms another hypothesis about the perceptual variable that might be under control and instigates another change. Another preferred state could be maintaining distance from a cell mate, which could be tested by a change of cell. The generation of alternative hypotheses would be most effective in collaboration with the whole team. This allows all members can use their varied perspectives to inform hypotheses of perceptual variables under control and generate ideas to test them. The use of psychological formulation with staff teams currently entails hypothesis generation (National Offender Management Service, 2015) but there is less emphasis on the iterative process of testing and updating hypotheses. This process places the emphasis on the practitioners to be creative about what perceptual variables might be controlled and how they can implement tests of these variables. The TCV is arguably much simpler to implement than traditional behaviour analysis and does not require long periods of behavioural recording. The approach is, however, dependent on practitioner's capacity to make changes to prisoners' environment or routine, which can be challenged by institutional settings where there are limited resources and inflexible policies.

More General Implications of PCT for Implementing Services

Other considerations regarding the way in which forensic practitioners implement interventions might also benefit by further examination from a PCT perspective. As has been explained, PCT understands perceived coercion to occur when a person controls one perceptual variable, such as completing a sentence plan, but this creates a state of conflict for them as it disturbs achievement of another of their goals. Conflict is a debilitating state because it reduces the effectiveness of the person's ability to control what they want. This suggestion is consistent with the finding noted earlier that mandated interventions are less effective than voluntary interventions (Parhar et al. 2008). As a mandated sentence disregards the goals and perspectives of the person being forced to attend the treatment, it is more likely to act as a disturbance to the achievement of their goals.

The concepts of mandated versus voluntary treatment are unlikely to be easily separated. Some individuals may complete an intervention regardless of the mandated requirement to do so. Those in voluntary treatment may still experience perceived coercion as individuals in their wider system – family, friends, or key workers – may be encouraging them to engage in an intervention. Another consideration is that interventions in criminal justice settings can be extremely long and/or residential. For example, the anger management intervention Reasoning and Rehabilitation can stretch to more than 36 two-hour sessions (Ross & Ross, 1995). However, we have not been able to identify any supporting research on why this duration of treatment is necessary for this difficulty. Even if this evidence did exist it is not clear whether this length of treatment is necessary for all people receiving it. In the psychotherapy field, there has been a long tradition of fixed duration therapies and yet there is an absence of evidence that a certain number of sessions are required for change to occur. Indeed, Barkham et al. (2006) have reported that treatment length is relatively unimportant for outcomes, with the overall picture being that patients can attend a wide variety of treatment durations to achieve the same amount of change. In a similar way it is possible that some participants may notice large improvements early on in an anger management program but have to continue with an extremely long program of therapy, potentially to the detriment of goals they may have. In this way, long treatment duration in criminal justice settings could be perceived by participants to be coercive, even if initial entry to the programme is voluntary. This is because the participant may perceive an improvement in their difficulties and their understanding of them.

It may be that some clients do not complete a course of therapy, and in this case may be labelled as treatment "dropouts". However, the notion of "dropping out" of therapy entails an assumption that a predetermined number of sessions that is necessary for change to occur, with shorter durations being premature, discontinued, or incomplete episodes of therapeutic activity.

Summary and Conclusions

Coercion is a concept that has attracted longstanding attention in the sociological and criminological literatures, but this attention has not elaborated specific detail of how coercion is manifest in interpersonal interactions and what mechanism underpins its emergence. We have described how behaviourist ideas can provide an elaborated description of reciprocal interactions when coercion is evident. However, they do not explain why attempts to use reinforcement to coerce people, and control behaviour, are so often countercontrolled by the people targeted by these strategies. We argue that this is because behaviourist ideas lack an appreciation that the purpose of behaviour is to achieve or maintain preferred perceptual states.

In the context of forensic practice each client or caseworker has his or her own set of perspectives on the rehabilitation or care that should occur and will act to ensure these states are achieved. Given that it is unlikely clients and practitioners will have the same definition, any intervention or management plan will inevitably disturb the states the other wishes to maintain. This is why it is essential for managers and services to attempt to learn as much as they can about the goals of the clients to avoid the possibility of counter control emerging.

In the current chapter we have also outlined how PCT explains why coercion is best described as a subjective – internal – state. Given that all behaviour is an outward component of control even when coerced, an individual is achieving *their own* goals required to satisfy other people. The problem arises when these goals conflict with other preferences they might have, and an unpleasant conflict emerges with coercion being experienced. Understanding what these conflicting goals are and how they might be achieved, without harming others, should be central to the delivery of effective and efficient care and rehabilitation of clients.

References

Angiolini, E. (2017). *Report of the Independent Review of Deaths and Serious Incidents in Police Custody*. United Kingdom Government Home Office.

Barkham, M. et al. (2006). Dose-effect relations and responsive regulation of treatment duration: the good enough level. *Journal of Consulting and Clinical Psychology*, 74, 160–167.

Bourbon, W. T. (1990). Invitation to the dance. *American Behavioral Scientist*, 34, 95–105.

Bourbon, W. T. (1995). Perceptual control theory. In H. L. Roitblat & J.-A. Meyer (eds), *Comparative approaches to cognitive science* (pp. 151–172). MIT Press/Bradford Books.

Braga, T., Cunha, O. & Maia, A. (2018). The enduring effect of maltreatment on antisocial behavior: A meta-analysis of longitudinal studies. *Aggression and Violent Behavior*, 40, 91–100.

Butterfill, S. (2012). Joint action and development. *Philosophy Quarterly*, 62(246), 23–47.

Carey, T. A. (2012). *Control in the classroom*. Hayward.

Carey, T. A. (2016). Beyond patient lefted care: Enhancing the patient experience in mental health services thorough patient-perspective care. *Patient Experience Journal*, 3, 46–49.

Carey. T. A. (2018). *Patient perspective care: A new paradigm for health systems and services*. Routledge.

Carey, T. A. & Bourbon, W. T. (2004). Countercontrol: A new look at some old problems. *Intervention in School and Clinic*, 40(1), 3–9.

Champion, M. K. (2007). Commentary: Seclusion and restraint in corrections – a time for change. *Journal of the American Academy Psychiatry and the Law*, 35, 426–430.

Colvin, M. (2000). *Crime and coercion*. St Martin's Press.

Colvin, M., Cullen, F. T., & Vander Ven, T. (2002). Coercion, social support, and crime: An emerging theoretical consensus. *Criminology*, 40, 19–42.

Coviello, D. M., Zanis, D. A., Wesnoski, S. A., Palman, N., Gur, A., Lynch, K. G. & McKay, J. R. (2013). Does mandating offenders to treatment improve completion rates? *Journal of Substance Abuse Treatment*, 44, 417–425.

Davidson, H. (2016). 'Abu Ghraib'-style images of children in detention in Australia trigger public inquiry. *Guardian*, 26 July.

Dishion, T. J., & Patterson, G. R. (2006). The development and ecology of antisocial behavior. In D. Cicchetti & D. J. Cohen (eds), *Developmental psychopathology, vol. 3: Risk, disorder, and adaptation*. Wiley.

Eyberg, S. M., Nelson, M. M., & Boggs, S. R. (2008). Evidence-based psychosocial treatments for children and adolescents with disruptive behavior. *Journal of Clinical Child and Adolescent Psychology*, 37, 215–237.

Independent Chief Inspector of Borders and Immigration. (2016). Annual report for the period 2015 to 2016. United Kingdom Government Home Office.

Liberty. (2018). A guide to the hostile environment. Retrieved from www.liberty humanrights.org.uk/sites/default/files/HE%20web.pdf

Marken, R. M. (1988). The nature of behavior: Control as fact and theory. *Behavioral Science*, 33, 196–206.

Marken, R. M. (1991). *Mind readings: Experimental studies of purpose*. Control Systems Group.

Marken, R. M., & Carey, T. A. (2015). *Controlling people: The paradoxical nature of being human*. Australian Academic Press.

Marquart, J. W. (1986). Prison guards and the use of physical coercion as a mechanism of prisoner control. *Criminology*, 24, 346–366.

Ministry of Justice. (2017). Safety in custody. *Quarterly Bulletin*, 2 August.

National Offender Management Service. (2015). *Working with offenders with personality disorder: a practitioners guide*. NHS England.

Newton-Howes, G., & Mullen, R. (2011). Coercion in psychiatric care: Systematic review of correlates and themes. *Psychiatric Services*, 62, 465–470.

Parhar, K. K., Wormith, S. J., Derkzen, D. D., & Beauregard, A. (2008). Offender coercion in treatment: A meta-analysis of effectiveness. *Criminal Justice & Behavior*, 35, 1109–1135.

Patterson, G. R. (1982). *A social learning approach, vol. 3: Coercive family process*. Castalia.

Powers, W. T. (1998). *Making sense of behavior*. Benchmark.

Powers, W. T. (1973). *Behavior: The control of perception*. Aldine.

Read, J., van Os, J., Morrison, A. P., & Ross, C. A. (2005). Childhood trauma, psychosis and schizophrenia: a literature review with theoretical and clinical implications. *Acta Psychiatrica Scandivica*, 112, 330–350.

Rogers, A., & Budd, M. (2016). Developing strong foundations: The DART approach. In A. Rogers, J. Harvey, & H. Law (eds), *Young people in forensic mental health settings: Psychological thinking and practice*. Palgrave Macmillan.

Ross, R. R., & Ross, R. D. (1995). The R&R programme. In R. R. Ross, & R. D. Ross (eds), *Thinking straight: The reasoning and rehabilitation programme for delinquency prevention and offender rehabilitation* (pp. 83–120). AIR Training and Publications.

Royal Commission to the North Territory Government. (2017). Protection and detention of children in the Northern Territory. Retrieved from https://child detentionnt.royalcommission.gov.au/Documents/Royal-Commission-NT-Report-O verview.pdf.

Runkel, P. (1990). Research method for control theory. *American Behavioral Scientist*, 34, 14–23.

Skinner, B. F. (1953). *Science and human behaviour*. Free Press.

Taylor, D. (2018). UK removed legal protection for Windrush immigrants in 2014. Retrieved from www.theguardian.com/uk-news/2018/apr/16/immigration-law-key-cla use-protecting-windrush-immigrants-removed-in-2014.

United Kingdom Home Office. (2019). *Modern slavery: A briefing.* Retrieved from https://assets.publishing.service.gov.uk/government/uploads/system/uploads/attachment_data/file/638368/MS_-_a_briefing_NCA_v2.pdf.

United Nations. (2013). *Prison incident management handbook.* United Nations.

9 Ostracism

Dennis Kaip and Joel Harvey

Prisons, and other secure settings, in essence involve social exclusion. The moment a person is handcuffed and placed in a cell under state control, there is a marked separation between them and free society. They are *of* the free world but not *in* the free world. Separation, and the sense of loss that it entails, has been acknowledged in classic sociological ethnographic studies (Sykes 1958), and continues to be central to the experience of imprisonment (Harvey, 2012). Imprisonment has been referred to as 'social death' (Price, 2015; Stearns et al., 2019). Ostracism- the process of being ignored and excluded by others (Williams & Nida, 2017) plays out at different levels for people experiencing a deprivation of liberty.

This chapter sets out to explore the importance of practitioners holding in mind the construct of ostracism when working with clients in forensic settings. We begin by exploring the concept of *ostracism* and the research that underpins it (Williams, 2007; Zadro et al., 2004). In particular, we examine the temporal need-threat model of ostracism (Williams, 2009) and its utility to forensic practice. We then explore the usefulness of explicitly thinking about the experience of ostracism when assessing, formulating, and carrying out psychological interventions with clients in forensic settings.

Belonging and Ostracism

Among the most basic and fundamental human needs is the need to belong (Baumeister & Leary, 1995; Miller, 2003). Yet, defining what 'belonging' means and what it encompasses is no easy feat, not least due to its socio-political implications, and the challenge of establishing a consensus on its definition and constituents (Miller, 2003). Nevertheless, belonging is widely understood as the process by which individuals develop a sense of identity within their social and emotional environment (ibid.). It is through interacting with others that a sense of self develops, and values, traditions, and collective norms are formed (May, 2011). Indeed, belonging connects the individual to the social (ibid.).

In contrast to belonging stands social exclusion. Social exclusion has broadly been defined as the experience of being kept apart from others,

DOI: 10.4324/9781315560243-9

either physically (e.g. social isolation) or emotionally (e.g. being ignored or explicitly told that one is unwanted) (Riva & Eck, 2016; Zadro et al., 2004).

Rejection-based experiences have been defined as those that "involve direct negative attention that conveys relational devaluation or otherwise indicates someone (or their group) is unwanted" (Wesselmann & Williams, 2017, p. 694). Examples of these types of experiences include numerous factors, such as being spoken to with dehumanising language or exclusive laughter, where people are at the 'receiving end' of a joke (Klages & Wirth, 2014; Wesselmann et al., 2018). It also includes the experience of micro-aggressions in the form of subtle, often automated, comments, and discriminatory behaviour such as being overlooked, under-respected, and devalued; a regular experience for many minority groups (Constantine, 2007).

Ostracism-based experiences have been defined as those that involve the person being ignored and excluded (Williams, 2009). These can include experiences in which individuals are *explicitly* ignored or discounted more subtly through the use of non-verbal cues (e.g. not being looked at,), uncomfortable silences, or being forgotten (Böckler et al., 2014; King & Geise, 2011; Koudenburg et al., 2011). Importantly, ostracism does not only apply on an interpersonal level, but also on a human organisational level, as research has demonstrated that the majority of people experience being ignored and excluded at least once a day (Hartgerink et al., 2015). This supplements the fact that ostracism impacts not just the ostracised person (intrapersonal effects), but others (interpersonal effects) as well (Hartgerink et al., 2015).

Research has demonstrated that ostracism can have negative effects on behaviour, personality, and social identity, as well as being a threat to fundamental human needs: belonging, self-esteem, control, and meaningful existence (Zadro et al., 2004). Consequences of ostracism can include an increase in aggression (Twenge et al., 2001), a reduction of pro-social behaviour and impaired self-regulation (Twenge et al., 2007).

Furthermore, while an ostracised individual can react to their experience in an antisocial manner, they can also respond to it in a pro-social manner (Wesselmann & Williams, 2017). Indeed, paradoxically, ostracised individuals can be more likely to conform, comply, and be obedient than their non-ostracised counterparts (Carter-Sowell et al., 2008; Riva et al., 2014; Wesselmann et al., 2018).

Ostracism can lead to negative emotional reactions, even when it is perpetrated by unrelated or disliked individuals and groups (Gonsalkorale & Williams, 2007; Williams et al., 2000; Zadro et al., 2004). Moreover, by posing a direct threat to basic psychological needs, ostracism can have negative implications on an individual's overall wellbeing, and lead to psychological difficulties such as anxiety and depression (Gausel & Thørrisen, 2014; Twenge et al., 2003; Wesselmann & Williams, 2017).

The Temporal Need-Threat Model of Ostracism

One model that has been developed to understand, in more detail, how individuals respond to ostracism, is the temporal need-threat model of ostracism (Williams, 2009). Fundamentally, Williams (2009) proposed that, following an experience of ostracism, an individual will attempt to either solve or cope with the now threatened basic need. This model postulates that following an initial pain response, an appraisal process is initiated that navigates the need-fortifying responses (Williams, 2009). Moreover, within this model, it is proposed that an individual can either be susceptible to social influence, or respond in a manner that asserts control, demands recognition, or is provocative or aggressive. Additionally, this model delineates between the short-term and long-term reactions to ostracism and proposed three stages. The stages are *reflexive*, *reflective*, and *resignation*.

Reflexive Stage

During the first stage, the individual is exposed to either a real or perceived experience of ostracism, and confronted with negative emotions, including anger. The response to the experience is often immediate due to the four fundamental needs being under attack: namely the need to belong, the need to maintain self-esteem, the need to perceive control over one's social environment, and the need to feel recognised for existing (Williams, 2009).

Reflective Stage

Following this initial 'reflexive reaction', the individual is deploying cognitive and behavioural tactics to work out why the ostracism occurred. Importantly, during this stage, the individual assesses and evaluates their experience of ostracism and the extent of the social injury. The intention is to make sense of the ostracism experience and respond to it in a manner that is deemed suitable to the ostracised person, depending on the need threatened.

In the reflective stage, the individual is devising methods to cope with, and respond to, the threatened need(s). In instances where the desire to belong or restore one's self-esteem is most threatened, a person might respond prosocially. Alternatively, if the need centres around regaining control or being seen to have a meaningful existence (i.e. to be noticed), a person may instead respond anti-socially.

During this stage, it is argued that it is the most threatened need that the individual is attempting to regain. In the coping phase, the person tries to re-secure their place in the group, attempting to regain the control that has been lost by lashing out in some way (Williams, 2009). Indeed, it is the case that "the most threatened needs direct the coping goals" (Williams & Nida, 2011, p. 72).

Resignation Stage

If, however, the ostracism continues, a prolonged sense of alienation, depression, helplessness, and worthlessness can occur (Riva et al., 2017). This is particularly pertinent within the prison population as individuals here are more likely to have experienced multi-layered rejection and ostracism. A sense of belonging and positive self-esteem is a threatened commodity. However, if the ostracised individual believes that re-inclusion and acceptance are possible, those needs will fuel the strive for re-inclusion, which leads to conformity.

Several studies have found support for the first two stages of this model (Williams et al., 2000; Zadro et al., 2004;) but, as Aureli et al. (2020) point out, there has been less attention paid to the effects of long-term ostracism and the resignation stage. However, the temporal need-threat model of ostracism is considered among the most suitable to investigate the underlying processes and reactions of an individual to ostracism (Jamieson et al., 2010).

Ostracism and Violence

It is of relevance to forensic practice to highlight that research within social psychology has consistently shown evidence for the relationship between ostracism and aggression (Reijntjes et al. 2011; Ren et al. 2018). Chow, Tiedens, and Govan (2008) examined the emotion of anger as a mediator of the relationship between ostracism and violence. Their results indicated that, in response to ostracism, anger had a greater predictive validity than any other negative emotion in anticipating an antisocial response. Additionally, when ostracism produced anger, it resulted in a greater expression of aggression.

One reason for an aggressive reaction is that ostracism can be a hurtful experience, and the person experiencing it might have learned that behaving aggressively alleviates those negative feelings (Bushman et al., 2001; Ren et al., 2018). For example, based on earlier life experiences, a prisoner might be prone to being easily slighted or shamed. Feeling a sense of shame by being ostracised in public could result in feelings of anger. Imagine a scenario in which a prisoner, out on a wing, asks another prisoner if they want to play a game of pool, but is then ignored. If the former is, in fact, hyper-vigilant to experiences of being slighted and shamed (due to earlier, negative life experiences), this scenario might lead to acute feelings of rejection, embarrassment, or shame. Feeling rejected, this individual might then assume that displaying aggression could improve, or even reverse, their negative, aversive experience (Chester & DeWall, 2017). Therefore, perceiving oneself as the target of ostracism may result in the elicitation of negative feelings.

Furthermore, acts of aggression can give an ostracised individual a false sense of control over a particular circumstance (Ren et al., 2018). Ren and colleagues proposed that aggression might be the preferred modus operandi for an individual if they experience a lack of control over a situation, specific circumstance, or even their existence. As ostracism can be a painful experience, responding aggressively can create a sense of relief from the situation for the individual. It may appear as a means to alleviate one's aversive state (Bushman et al., 2001; Chester & DeWall, 2017), and a tool to regain control.

One particular form of violence that, among other factors, has been related to ostracism, is school shootings. Leary et al. (2003) conducted a case study of 15 school shootings between 1995 and 2001, and found that, in addition to other factors (e.g. the experience of bullying), ostracism was present in all except two school shooting events. Moreover, a systematic review by Sommer, Leuschner, and Scheithauer (2014) analysed 126 school shooting cases from 13 countries and found that some variation of peer rejection or exclusion accounted for about 70 per cent of the school shooting cases.

Other research has examined the association between the experience of ostracism and violent extremism (Hales & Williams, 2019; Knapton, 2014). Through a review of the literature, Knapton (2014) concluded that individuals who experience ostracism could have an urge to strengthen (defend) the areas of their emotional wellbeing that they feel are targeted by ostracism (e.g. their feeling of self-esteem or sense of belonging). As a result, they may bond with other individuals who have, or have had, similar experiences. Indeed, other research has shown that ostracised individuals are more likely to conform with a norm, comply with a request, and show greater obedience towards an authority figure, than individuals who have not felt ostracised, or have not had experiences that led them to feel ostracism (Carter-Sowell et al., 2008; Riva, et al., 2014; Williams et al., 2000). Such dynamics could be at play within the process of radicalisation.

Importantly, the selectivity and exclusivity that extremist groups portray, of who can and cannot join their group, can be tempting to an ostracised individual as this might provide a greater sense of acceptance, and an increase in self-esteem (Hales & Williams, 2018). Moreover, research posits that the extreme and violent actions of those groups might also elicit a sense of control, eliminates perceived uncertainty, and forces acknowledgement from the outside (Hales et al., 2018; Williams et al., 2019). Concerningly, Pfundmair (2019) proposed a direct link between an ostracism-induced threat to basic psychological needs and the motivation for terroristic aggression.

Ostracism and the Experience of Imprisonment

We now turn to consider the role of ostracism within the experience of imprisonment. As shown in Figure 9.1, the experience of ostracism for prisoners plays out at a number of different levels. Firstly, before even arriving at a prison or secure hospital, the majority of individuals have already experienced chronic social exclusion and rejection in their lives (Kyprianides et al., 2019). It is well established that people who have contact with the criminal justice system are also more likely to be homeless, have been excluded from school, and experienced disrupted attachments and

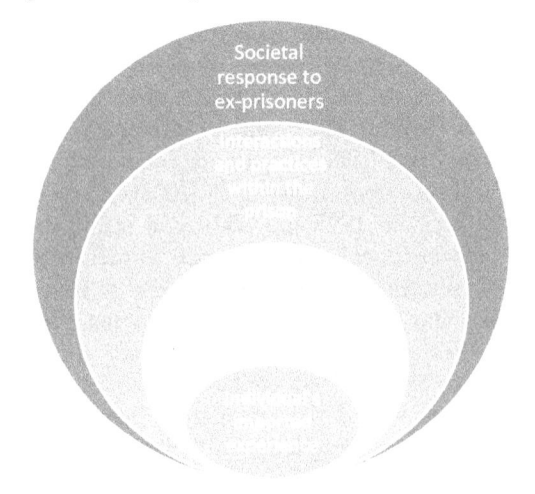

Figure 9.1 Multi-layered ostracism

trauma (Messina & Schepps, 2021; Social Exclusion Unit, 2002). Moreover, it has been shown that a large number of individuals, both male and female, within a prison population have a significantly greater number of Adverse Childhood Experiences (ACE) than the general population (Messina & Schepps, 2021). Originally, Felitti et al. (1998) identified ten specific negative exposures that could predict negative health outcomes: physical abuse, emotional abuse, sexual abuse, physical neglect, emotional neglect, household substance abuse, violent treatment towards mother, parental separation or divorce, household mental illness, and having a household member incarcerated. We would argue that the experience of neglect and separation can be experienced as ostracism by family and others.

Secondly, the mere act of being sent to prison signifies that the individual has been separated and excluded from mainstream society. An individual in a secure forensic setting goes on to experience the loss of many of their rights and social benefits, in tandem with the more general experience of being excluded from society for the period of their incarceration (Balafoutas et al., 2020). On a social level, prisons parallel the interpersonal phenomenon of punitive ostracism (Aureli et al., 2020). The separation from, and loss of, family and friends is central here.

Thirdly, within the prison walls, ostracism can occur through interactions and practices. The prison's internal structures and culture can add further layers of ostracism onto what the individual has already experienced. For example, some in-prison strategies, such as solitary confinement, mechanise ostracism in order to bring about discipline and control (Arrigo & Bullock, 2008; Haney, 2018). By virtue of depriving an individual of their fundamental human need for connection, this practice can have severe, detrimental consequences on one's mental wellbeing.

The incarcerated individual can also be excluded by fellow prisoners or correctional staff (Aureli et al., 2020). However, while there has been extensive research into prisoners' experience of bullying (Allison & Ireland, 2010; Gavin, 2019; Ireland, 2014; South & Wood, 2006), there has been less of a focus on examining the impact of ostracism from fellow prisoners or staff. For example, what impact does it have on a prisoner who is pressing their cell bell to get a response from staff, but is not attended to? What impact does it have on a prisoner when their cell mate does not speak to them? What is the psychological impact of being excluded from discussions when interacting with other prisoners on the exercise yard?

Aureli et al. (2020) did examine the effect of ostracism on prisoners, and the importance and impact of group support. They postulated that the aversive consequences of socially excluding an already frequently excluded cohort could be moderated through group-based intervention. To test their hypothesis, they examined the effects of social exclusion by comparing a group of 68 prisoners (31 of whom engaged in support groups) with 68 men in the general population, in Italy. Results demonstrated that prisoners who did not have access to a support group showed greater levels of feelings of resignation. Conversely, in terms of resignation outcomes, prisoners who engaged in a support group had the same results as the non-prisoner cohort.

Fourthly, prison-based ostracism does not necessarily end when the individual leaves the gates. Upon release, the ex-prisoner comes face-to-face with several obstacles that challenge successful reintegration into society (Musa & Ahmed, 2015). For example, through examining the experiences of 199 ex-prisoners, Kyprianides et al. (2019) found that individuals continue to experience ostracism as a result of their previous incarceration. Among other things, their results showed that identifying as an ex-prisoner amplified the relationship between social rejection and reduced overall well-being.

While discrimination based on race, ethnicity, and gender is against the law in many countries, there are often no laws in place that provide equal or similar protection against discrimination towards ex-offenders or ex-prisoners. In some instances, such discriminations are even considered legitimate (Kyprianides et al., 2019). Moreover, society often harbours stigmatising attitudes towards ex-prisoners which makes reintegration into the community following release a difficult endeavour (Hirschfield & Piquero, 2010). As a result, ex-prisoners experience difficulty when trying to find housing or employment and do so while facing the stigmatising judgements and attitudes from society, housing providers, and employers. In turn, this may have a deep and negative impact on the success of their reintegration into society.

Finally, it is important to highlight that the experience of ostracism is not just limited to the prisoner or ex-prisoner, but often extends to their families (Murray, 2007; Murray & Farrington, 2008; Musa & Ahmed, 2015). Parental incarceration, for example, can have a myriad of detrimental consequences for children and the non-imprisoned parent. The imprisoned parent may have played a fundamental role in providing care, emotional support, and education to the child, as well as embodied an important authority figure in the child's life. In addition, they could be responsible for, or even the cause of, financial difficulties, resulting in loss of housing and/or a lack of food or basic items.

Additionally, by complicating the roles of caregiving and authority, and potentially increasing a child's susceptibility to experiencing the negative effects of societal stigma, parental imprisonment can lead to family break-ups (Lanskey et al., 2019; Murray & Farrington, 2008; Robertson et al., 2016). For example, imprisoned women may have their children taken into local authority care. Parental imprisonment is a traumatic experience that can increase the risk of mental health problems, antisocial behaviour, substance misuse, and offending, and can ultimately lead to education and employment problems (Murray & Farrington, 2008). Further concerning consequences of parental imprisonment were found by Van de Weijer et al. (2018) who examined the difference between the effects of having incarcerated parents versus parents who engaged in criminality but were not incarcerated. Their results indicated that children of incarcerated parents are at a greater risk of dying prematurely, compared to those with either non-offending parents, and/or offending parents with a criminal background but who were not incarcerated. This implies that it is parental imprisonment, not parental criminality, that increases the risk of mortality. In turn, a cycle of social exclusion can occur, and it is noted that children of imprisoned parents are the 'forgotten victims' within the criminal justice system (Matthews, 1983, as cited in Murray & Farrington, 2008).

Ostracism, Social Death, and Suicidality in Prisons

When prisoners enter prison, it is argued that they experience a form of 'social death'. This term offers a powerful metaphor to describe the social consequences of ostracism (Bauman, 1992; Williams, 2007). The precise definition of the term is still debated among scholars (Králová, 2015) but it is widely agreed that 'social death' occurs when an individual or group experiences a severe or substantial loss, such as social exclusion or the loss of one's social identity (Králová, 2015; Steele et al., 2015).

Our connections and sense of belonging are of fundamental importance to our psychological and physiological well-being and survival (Baumeister & Leary, 1995; Cacioppo & Patrick, 2008; Diener & Seligman, 2002; Heine et al., 2006; Waytz & Epley, 2012). The experience of rejection elicits brain activity that is associated with processing physical pain (Eisenberger et al., 2003; Steele et al., 2015). There is also a plethora of empirical studies that allude to the link between lethal suicidal behaviour and numerous layers of isolation, such as social withdrawal, lack of social support, loneliness, and/or residing in a single prison cell (Fazel et al., 2008; Nickel et al., 2006; Wyder et al., 2009). Indeed, it is important for forensic practitioners to consider how the experience of imprisonment, and the 'social death' that might ensue, has relevance to the assessment of suicidality.

Harvey and Liebling (2001) put forward a conceptual model that explores how an individual could spiral towards suicidal ideation and behaviour while in prison. They argue that, at the interpersonal level, ostracism can influence an individual's ability to cope, which in turn can increase the likelihood of suicidal behaviour. The model explores how a lack of social support and ostracism can act as a catalyst for an acceleration towards a 'suicidal threshold', and how suicidality can be the direct consequence of prolonged ostracism. If an individual remains anchored within this spiral, the risk of 'social death' is increasingly pronounced. Importantly, it is postulated that if social support is offered to an individual after they have moved down the spiral to a state of hopelessness and helplessness, it might not be accepted as the individual in need might find it difficult to respond to what is offered. Within that model, the institutional response and potential contribution to suicidal behaviour is acknowledged.

A mediating factor to further consider is that of interpersonal rejection sensitivity and its relationship with both aggression and suicidality. Interpersonal rejection sensitivity is understood as "a disposition to anxiously expect, readily perceive, and overreact to rejection" (Downey & Feldman, 1996, p. 1338). Indeed, in a study by Williams, Doorley and Esposito-Smythers (2017) on eighty psychiatrically hospitalised adolescents and their families, it was found that interpersonal rejection sensitivity mediated the association between peer victimisation and severity of both aggression and suicidal ideation. This is complimented by further research which proposes that social rejection is fundamentally involved in the development of suicidal ideation (Brown et al., 2019; Campos & Holden, 2015; Mitchell et al., 2018).

Clinical Implications

Having explored the concept of ostracism, and how it potentially plays out in relation to the prisoner experience, we will now explore its clinical utility. We argue that when working clinically with prisoners in forensic settings, it is important to consider the role of ostracism in relation to the prisoner experience and in relation to the risk of harm to self and others. To illustrate this, we start by introducing a composite case vignette of Adam, a young man in his mid-twenties from the Irish traveller community, and we apply the temporal need-threat model of ostracism (Williams, 2009).

Vignette

The following vignette presents information that was obtained from an initial assessment with 'Adam'. As can be seen, the assessment found that Adam has experienced significant and cumulative ostracism, from multiple sources and in multiple contexts, throughout his life.

Adam was referred to the mental health team from the prison officer team. The referral stated that he reported hearing voices and having difficulty sleeping due to violent flashbacks of his aggression towards prison officers and inmates on the landings. Adam's past offences include aggravated assault and burglary. Most recently, he was charged with Grievous Bodily Harm (GBH) and awaiting a potentially long prison sentence.

Following an initial meeting with the wing staff, it became evident that many of the more inexperienced officers felt intimated by Adam's regular displays of aggression, while the more experienced staff appeared to feel resentment towards him for the injuries that he inflicted on some of their colleagues. Most prison officers considered him dangerous and erratic. Additionally, there seemed to be a strong consensus among the uniformed staff that Adam should spend as little time outside of his cell as possible. However, Adam was holding a job on the landing and believed to be involved in dealing substances within the prison. Adam's status as a worker led to tension among the uniformed staff on the wing and him being treated inconsistently. When enquiring about any further observations about Adam, the feedback given by the attending officers suggested that he "is just a gypsy, they're all mad anyway", that "he's fine, they don't talk to us", and "they're all lying so why bother?"

Early reports showed that, while Adam moved in and out of foster care during his childhood, his biological parents also became involved with the criminal justice system (CJS). Additionally, most of his placements ended up being terminated as he repeatedly absconded. Due to a breakdown of multiple foster care placements, Adam was eventually sent to several secure care homes across the country. It was there where he became a witness and victim to violence from other young people as well as staff. Adam's first contact with the CJS was at the age of 13 when he broke into a house. He was eventually sent to a Young Offender Institution (YOI) at the age of 15. While Adam was in and out of institutions for most of his adult life, he never once had an experience with a mental health service, even when he was in prison or out in the community.

Adam shared with an officer that he learned to fight "the day I learned to walk", and that a feeling of distrust towards any outsiders was instilled in him and his siblings. He explained that his girlfriend came from the same community but had mental health difficulties. Their relationship was especially tumultuous during her pregnancy, because, as well as during the birth of his child, Adam was in prison. Subsequently, he has had no contact with his girlfriend or child other than through pictures or phone calls which stopped after a while. Adam also explained that he was part of a gang, and that he had conflict with rival gangs.

Adam explained that it was not common to seek support for one's own mental health within the travelling community, as this was regarded as a weakness or even a form of betrayal towards one's family. He added that it was believed that it is one's family who is supposed to help you, and not "outsiders, who don't understand our ways". However, he shared that he had difficult relationships with members of his family, and that many of them did not want to speak with him. He did not want to provide details about why this was the case and said nobody would understand.

Initially, Adam seemed eager to present as a strong, combat-ready man who can fend for himself without having to rely on outsiders "who wouldn't help anyways". Adam began to report that he was very young when he was taken away from his family. He shared that nobody in his family seemed to fight for him or want to keep him. However, his goal was to fight to get his own family back on his side, as well as solving matters with his partner to have his own family taken care of. Adam explained that the relationships with some members of his family were complicated but that he had a cousin who stayed in contact with him throughout his time in prison, and who had promised to get him a job in construction. However, Adam carefully added that he was unsure as to whether this would actually happen, but that he had no other choice than to accept any support coming from within his family, especially as "outsiders don't give jobs to travellers".

He explained that most of his problems with outsiders began when his foster carers forced him to call them "Mum and Dad", and insisted that he should not be sad about not being with his "kind" anymore. Throughout his time in different foster care placements, he was told that the traveller community was not to be trusted, and that he should consider himself lucky that he had been taken away. However, Adam explained that he refused to give up on his family and his traditions- a defiance that led to him getting punished. This was one of the main reasons why he started to run away from his foster placement, which resulted in the police being involved in his life again, as well as placement breakdowns, and being sent to a residential home. It was here where Adam often experienced bullying from other young people and even members of staff who would regularly call him the "crazy gypsy", and punish him more harshly than the other young people.

Adam shared that it was his time in care that made him realise that if he did not stand up for himself, nobody else would. If he was the "last man standing", nothing and nobody could control him. When he ended up in a YOI, he learned that you needed to be quick to build a reputation of someone who better not to be messed with.

The event of a rival gang breaking into his home and assaulting him brought back memories from his childhood of when a male relative would come into his room at night, drunk, and sexually assault him. Though Adam did not reveal any further details of these experiences, he shared that he labelled them as 'visits from the weirdo' which occurred frequently throughout his childhood. While speaking about these 'visits', Adam appeared visibly smaller, physically, and was clenching his fist. However, Adam promptly added that he believed that that person should not be blamed for being a drunk, because it was "the outside world that made the traveling man a horrible drinker".

Adam went on to explain that, from the moment he became a man and learned to use his fists, "nobody would come for visits anymore" or would dare to violate his personal space. This feeling of being powerful and able to protect himself changed the moment the rival gang broke into his home. Even though he was the one being attacked in his own home, and argued to have acted in self-defence, he was the one who was sent to prison, while those who attacked him evaded incarceration due to a poor pursuit by the police. This made Adam feel as though he was not treated fairly by the system and given the same chance as everyone else, reinforcing his distrust in the system as a whole.

Even now in prison, Adam shared that he feels like he is being treated differently because of his background. He shared circumstances in which he believed that he was sent to the 'seg' (segregation unit) much more quickly and for lesser infringements than other inmates who were not from the same community. He was convinced that the prison officers were acting out of bias because of his background and made him responsible for all of the problems on the wing. Adam felt that this was something that has not changed since his days in residential care and thought that his only option was to make sure he did not appear as weak, to avoid being taken advantage of.

It is important to note that, in addition to Adam's difficulties in trusting those outside of his community, the long-term sequalae of his ACEs add to his mistrust towards members of the mental health team. A study by Munoz et al. (2019) found that just one adverse experience during childhood can lead to less trust in the medical profession. This generally aligns with the theory of attachment which posits that trauma perpetrated by adults towards children can lead to less trust towards authority figures during adulthood (ibid.).

Application of the Model

Having obtained information from the initial assessment with Adam, the psychological formulation of his aggression can be situated within the temporal need-threat model of ostracism.

REFLEXIVE STAGE

Adam has been exposed to both real and perceived ostracism since early childhood. He learned early on that, because of his background, he would be treated differently. His emotional reactions, which ranged from sadness and jealousy to anger and hate, were responses to his fundamental needs being threatened; an experience he had endured since childhood. He was taken from his family, who did not seem to fight for him, and was placed in foster care where he was forced to abandon his identity. During his time in residential care, he was considered an outsider, and experienced other young people teaming up against him, and the unfairness of residential staff punishing him more severely than his peers. His self-esteem and sense of belonging did not have a chance to build and develop, as they were constantly under attack.

REFLECTIVE STAGE

Following Adam's 'reflexive reaction', and the severity of the social injury he endured throughout his life, he employed mostly anti-social methods to defend those threatened fundamental needs. Adam also seems to believe that he is responding appropriately to those who ostracised him and his community. He even feels as though the negative acts committed against him by his own family are justified- believing that they were merely a result of the negative treatment his family received from the outside community.

Moreover, Adam's way of coping is two-fold. Firstly, he believes that he must pose and appear as a threatening individual to avoid being taken advantage of. Secondly, having internalised that he cannot, and will not, be accepted by society due to his cultural background, Adam justifies his aggression as being a necessity to survive and avoid further social injury. He explains that being the aggressor first serves as a form of protection against those he believes wish to harm him, as it acts as a warning against any potential threat. His most threatened needs, his sense of belonging, and his sense of self, determine his response to a perceived threat. By believing to be in control, Adam tries to re-secure his place within his community, and regain his status as a valued member of his family.

RESIGNATION STAGE

As Adam's experience of ostracism continued from childhood to adolescence, and now in prison during his adulthood, he feels a sense of alienation, and shows clear signs of mental health difficulties. As previously mentioned, within the prison population, prisoners experience multi-layers of rejection and ostracism. Indeed, Adam's current situation demonstrates that this is, in fact, still the case and remains a current experience. His strained relationship with the prison officer team further reinforces Adam's opinions of himself, and functions as a subjective confirmation that his fundamental needs remain under threat. As previously stated, if the ostracised individual sees those needs being threatened, the possibility of re-inclusion becomes less likely.

Intervention

Given Adam's history, and based on the formulation of his difficulties, an intervention that could help him better understand how his life history has affected his interpersonal relationships with others, is Schema Therapy (ST; Young et al., 2003). ST is an integrative therapy that amalgamates concepts and theoretical formulations from various therapies, including cognitive (behavioural) therapies, Gestalt therapy, psychodynamic methods, and experiential theories (Keulen-de Vos et al., 2016; Young et al., 2003). Initially, ST was designed for, and aimed at, supporting individuals with chronic Axis I and Axis II disorders (Young et al., 2003). The foundation of ST is built around four main concepts, namely 'core emotional needs', 'early maladaptive schemas', 'schema modes', and 'maladaptive coping styles' (Young et al., 2003). ST aims to help individuals identify previously unmet emotional needs, and learn adaptive techniques to meet their needs by changing past cognitive, behavioural, and emotional patterns (Young et al., 2003). The role of social ostracism in moulding schemas is of relevance here.

Based on Bowlby's (1977) attachment theory, ST postulates that, with some variation, all individuals have core emotional needs which emerge from childhood (Young et al., 2003). These include the need for secure attachment, autonomy, realistic limitations, self-directedness, and spontaneity/play. The child's temperament, in tandem with the child's experiences in their environment, can lead to their core emotional needs being threatened or unmet. Thus, potentially resulting in unhealthy functioning and vulnerabilities that can sustain into adulthood.

Another pillar of ST is the concept of early maladaptive schema (EMS), which Young and colleagues define as a "broad, pervasive theme or pattern, comprised of memories, emotions, cognitions, and bodily sensations, regarding oneself and one's relationships with others, developed during childhood or adolescence, elaborated throughout one's lifetime and dysfunctional to a significant degree" (Young et al., 2003, p. 7). EMS can arise through the interaction between a child's emotional temperament (or personality) and (adverse) childhood experiences, and where one or more core emotional needs are not satisfied (ibid.).

Within the ST model, 18 EMSs are identified and grouped into four domains: (i) disconnection and rejection, (ii) impaired autonomy and performance, (iii) excessive responsibility and standards, and (iv) impaired limits (Lockwood & Perris, 2012, as cited in Bach et al., 2018). When an EMS is triggered, the person might employ maladaptive coping styles to avoid the unwanted experiences, or overcompensate by becoming aggressive (Derby et al., 2016; Pozza et al., 2020).

Another scaffold of ST is the concept of schema modes which control an individual's cognition, emotion, and behaviour (Young et al., 2003). Schema modes are defined as adaptive or maladaptive, that are currently active for an individual. A dysfunctional schema mode is activated when specific maladaptive schemas or coping responses have erupted into distressing emotions, avoidance responses, or self-defeating behaviours that take over and control an individual's functioning.

(Young et al., 2003, p. 37)

While some schema modes might be inactive, activation can occur at any stage, and the individual can switch between different modes (Young et al., 2003). Individuals who have been diagnosed with severe personality disorders might not have access to a 'healthy adult mode' that can balance the emotional states, thus leading to constant changes between modes (Bernstein et al., 2007; Young et al., 2003). Bernstein et al. (2007) argue that schema modes are tied to an individual's risk of violence and recidivism.

When examining Adam's life history, maladaptive EMS and schema modes appear evident. His chronic experiences of harassment and social exclusion from mainstream society will very likely have resulted in Adam developing a social isolation schema. This may manifest itself in the feeling that he does not belong anywhere, and the experience of a chronic sense of loneliness and detachment from others. Additionally, experiences of abuse/discrimination may have been internalised, hence why Adam is likely to be extremely self-critical and find it very difficult to trust others. The combination of social isolation, mistrust/abuse schemas, and an internalised self-critical voice, will have made connecting with others very difficult. As a way to navigate this pain and distress in order to survive, Adam seems to have developed some unhealthy schema modes. For example, he may have developed a mode that is hyper-vigilant to danger, and therefore attacks others before they can attack or hurt him. Additionally, he may have attempted to protect himself by keeping others at a distance, or cutting himself off from his own feelings. The role of ostracism and rejection throughout his life has contributed to the development of these difficulties.

It is important to help Adam understand these modes of behaving in the world, and the impact that they have had on his psychological health, his connection to others, and his risk of violence. The goal is to support Adam in moving away from the less adaptive sides of himself, and towards building healthier aspects of his personality, reducing his internal self-criticism, connecting to his vulnerabilities, and building emotionally connected relationships with others. The aim is to help reduce the risk of violence and decrease his level of psychological distress.

Conclusion

This chapter has explored the concept of ostracism within the context of forensic practice. Indeed, we have argued for the inclusion of social exclusion in forensic practice. Many prisoners import with them multiple experiences of social exclusion; experiences which then interact with an environment that might serve to exacerbate them further. While prison may sadly be a place of belonging for some prisoners, it is inherently an environment that is organised to separate and reject a group of people from mainstream society. Taking an ostracism lens to forensic practice shines a spotlight on the importance that everyday social interactions between practitioners and clients may have in helping breakdown feelings of social exclusion. Furthermore, it is important to continue considering the role of language in the context of practice, as it could further contribute to the experience of ostracism for clients, or indeed help them to develop a greater sense of feeling included in everyday life. It is important that forensic practitioners reflect on the language that they use when talking or writing about, and/or interacting with, a client, as it has the potential to either increase exclusion and develop 'otherness', or, more adaptively, increase inclusion and break down the division between 'us' and 'them'.

References

Allison, M. D., & Ireland, J. L. (2010). Staff and prisoner perceptions of physical and social environmental factors thought to be supportive of bullying: The role of bullying and fear of bullying. *International Journal of Law and Psychiatry*, 33(1), 43–51.

Arrigo, B. A., & Bullock, J. L. (2008). The psychological effects of solitary confinement on prisoners in supermax units: Reviewing what we know and recommending what should change. *International Journal of Offender Therapy and Comparative Criminology*, 52(6), 622–640.

Aureli, N., Marinucci, M., & Riva, P. (2020). Can the chronic exclusion-resignation link be broken? An analysis of support groups within prisons. *Journal of Applied Social Psychology*, 50(11), 638–650.

Bach, B., Lockwood, G., & Young, J. E. (2018). A new look at the schema therapy model: organization and role of early maladaptive schemas. *Cognitive Behaviour Therapy*, 47(4), 328–349.

Balafoutas, L., García-Gallego, A., Georgantzis, N., Jaber-Lopez, T., & Mitrokostas, E. (2020). Rehabilitation and social behavior: Experiments in prison. *Games and Economic Behavior*, 119, 148–171.

Bauman, Z. (1992). *Mortality, immortality and other life strategies*. Stanford University Press.

Baumeister, R. F., & Leary, M. R. (1995). The need to belong: desire for interpersonal attachments as a fundamental human motivation. *Psychological Bulletin*, 117(3), 497.

Bernstein, D. P., Arntz, A., & Vos, M. D. (2007). Schema focused therapy in forensic settings: Theoretical model and recommendations for best clinical practice. *International Journal of Forensic Mental Health*, 6(2), 169–183.

Böckler, A., Hömke, P., & Sebanz, N. (2014). Invisible man: Exclusion from shared attention affects gaze behavior and self-reports. *Social Psychological and Personality Science*, 5(2), 140–148.

Bowlby, J. (1977). The making and breaking of affectional bonds: I. Aetiology and psychopathology in the light of attachment theory. *The British Journal of Psychiatry*, 130(3), 201–210.

Brown, S. L., Mitchell, S. M., Roush, J. F., La Rosa, N. L., & Cukrowicz, K. C. (2019). Rejection sensitivity and suicide ideation among psychiatric inpatients: An integration of two theoretical models. *Psychiatry Research*, 272, 54–60.
Bushman, B. J., Baumeister, R. F., & Phillips, C. M. (2001). Do people aggress to improve their mood? Catharsis beliefs, affect regulation opportunity, and aggressive responding. *Journal of Personality and Social Psychology*, 81(1), 17.
Cacioppo, J. T., & Patrick, W. (2008). *Loneliness: Human nature and the need for social connection*. W. W. Norton & Company.
Campos, R. C., & Holden, R. R. (2015). Testing models relating rejection, depression, interpersonal needs, and psychache to suicide risk in nonclinical individuals. *Journal of Clinical Psychology*, 71(10), 994–1003.
Carter-Sowell, A. R., Chen, Z., & Williams, K. D. (2008). Ostracism increases social susceptibility. *Social Influence*, 3(3), 143–153.
Chester, D. S., & DeWall, C. N. (2017). Combating the sting of rejection with the pleasure of revenge: A new look at how emotion shapes aggression. *Journal of Personality and Social Psychology*, 112(3), 413.
Chow, R. M., Tiedens, L. Z., & Govan, C. L. (2008). Excluded emotions: The role of anger in antisocial responses to ostracism. *Journal of Experimental Social Psychology*, 44(3), 896–903.
Constantine, M. G. (2007). Racial microaggressions against African American clients in cross-racial counseling relationships. *Journal of Counseling Psychology*, 54(1), 1.
Derby, D. S., Peleg-Sagy, T., & Doron, G. (2016). Schema therapy in sex therapy: A theoretical conceptualization. *Journal of Sex & Marital Therapy*, 42(7), 648–658.
Diener, E., & Seligman, M. E. (2002). Very happy people. *Psychological Science*, 13(1), 81–84.
Downey, G., & Feldman, S. I. (1996). Implications of rejection sensitivity for intimate relationships. *Journal of Personality and Social Psychology*, 70(6), 1327.
Eisenberger, N. I., Lieberman, M. D., & Williams, K. D. (2003). Does rejection hurt? An fMRI study of social exclusion. *Science*, 302(5643), 290–292.
Fazel, S., Cartwright, J., Norman-Nott, A., & Hawton, K. (2008). Suicide in prisoners: a systematic review of risk factors. *The Journal of Clinical Psychiatry*, 69(11), 1721–1731.
Felitti, V. J., Anda, R. F., Nordenberg, D., Williamson, D. F., Spitz, A. M., Edwards, V., & Marks, J. S. (1998). Relationship of childhood abuse and household dysfunction to many of the leading causes of death in adults: The Adverse Childhood Experiences (ACE) Study. *American Journal of Preventive Medicine*, 14(4), 245–258.
Gausel, N., & Thørrisen, M. M. (2014). A theoretical model of multiple stigma: ostracized for being an inmate with intellectual disabilities. *Journal of Scandinavian Studies in Criminology and Crime Prevention*, 15(1), 89–95.
Gavin, P. (2019). 'Prison is the worst place a Traveller could be': the experiences of Irish Travellers in prison in England and Wales. *Irish Probation Journal*, 16, 135–152.
Gonsalkorale, K., & Williams, K. D. (2007). The KKK won't let me play: Ostracism even by a despised outgroup hurts. *European Journal of Social Psychology*, 37(6), 1176–1186.
Hartgerink, C. H., Van Beest, I., Wicherts, J. M., & Williams, K. D. (2015). The ordinal effects of ostracism: A meta-analysis of 120 Cyberball studies. *PloS One*, 10(5), e0127002.
Hales, A. H., & Williams, K. D. (2018). Marginalized individuals and extremism: The role of ostracism in openness to extreme groups. *Journal of Social Issues*, 74(1), 75–92.
Hales, A. H., & Williams, K. D. (2019). Extremism leads to ostracism. *Social Psychology*, 51, 149–156.

Haney, C. (2018). Restricting the use of solitary confinement. *Annual Review of Criminology*, 1, 285–310.

Harvey, J., & Liebling, A. (2001). Suicide et tentatives de suicide en prison: vulnérabilité, ostracisme et soutien social. *Criminologie*, 34(2), 57–83.

Harvey, J. (2012). *Young men in prison: Surviving and adapting to life inside*. Routledge.

Heine, S. J., Proulx, T., & Vohs, K. D. (2006). The meaning maintenance model: On the coherence of social motivations. *Personality and Social Psychology Review*, 10(2), 88–110.

Hirschfield, P. J., & Piquero, A. R. (2010). Normalization and legitimation: Modeling stigmatizing attitudes toward ex-offenders. *Criminology*, 48(1), 27–55.

Ireland, J. L. (2014). *Bullying among prisoners: Evidence, research and intervention strategies*. Routledge.

Jamieson, J. P., Harkins, S. G., & Williams, K. D. (2010). Need threat can motivate performance after ostracism. *Personality and Social Psychology Bulletin*, 36(5), 690–702.

Keulen-de Vos, M. E., Bernstein, D. P., Vanstipelen, S., de Vogel, V., Lucker, T. P., Slaats, M., Hartkoorn, M., & Arntz, A. (2016). Schema modes in criminal and violent behaviour of forensic cluster B PD patients: A retrospective and prospective study. *Legal and Criminological Psychology*, 21(1), 56–76.

King, L. A., & Geise, A. C. (2011). Being forgotten: Implications for the experience of meaning in life. *The Journal of Social Psychology*, 151(6), 696–709.

Klages, S. V., & Wirth, J. H. (2014). Excluded by laughter: Laughing until it hurts someone else. *The Journal of Social Psychology*, 154(1), 8–13.

Knapton, H. (2014). The recruitment and radicalisation of Western citizens: does ostracism have a role in homegrown terrorism? *Journal of European Psychology Students*, 5(1), 38–48.

Koudenburg, N., Postmes, T., & Gordijn, E. H. (2011). Disrupting the flow: How brief silences in group conversations affect social needs. *Journal of Experimental Social Psychology*, 47(2), 512–515.

Králová, J. (2015). What is social death? *Contemporary Social Science*, 10(3), 235–248.

Kyprianides, A., Easterbrook, M. J., & Cruwys, T. (2019). "I changed and hid my old ways": How social rejection and social identities shape well-being among ex-prisoners. *Journal of Applied Social Psychology*, 49(5), 283–294.

Lanskey, C., Markson, L., Souza, K., & Lösel, F. (2019). Prisoners' families' research: Developments, debates and directions. In M. Hutton & D. Moran (eds), *The Palgrave handbook of prison and the family* (pp. 15–40). Palgrave Macmillan.

Leary, M. R., Kowalski, R. M., Smith, L., & Phillips, S. (2003). Teasing, rejection, and violence: Case studies of the school shootings. *Aggressive Behavior: Official Journal of the International Society for Research on Aggression*, 29(3), 202–214.

Lockwood, G., & Perris, P. (2012). A new look at core emotional needs. In M. van Vreeswijk, J. Broersen, & M. Nadort (eds), *The Wiley-Blackwell handbook of schema therapy: Theory, research, and practice* (pp. 41–66). Wiley-Blackwell.

Matthews, J. (1983). *Forgotten victims: How prison affects the family*. National Association for the Care and Resettlement of Offenders.

May, V. (2011). Self, belonging and social change. *Sociology*, 45(3), 363–378.

Messina, N. P., & Schepps, M. (2021). Opening the proverbial 'can of worms' on trauma-specific treatment in prison: The association of adverse childhood experiences to treatment outcomes. *Clinical Psychology & Psychotherapy*, 28(5), 1210–1221.

Miller, L. (2003). Belonging to country—a philosophical anthropology. *Journal of Australian Studies*, 27(76), 215–223.

Mitchell, S. M., Seegan, P. L., Roush, J. F., Brown, S. L., Sustaíta, M. A., & Cukrowicz, K. C. (2018). Retrospective cyberbullying and suicide ideation: The mediating roles of depressive symptoms, perceived burdensomeness, and thwarted belongingness. *Journal of Interpersonal Violence*, 33(16), 2602–2620.

Munoz, R. T., Hanks, H., Brahm, N. C., Miller, C. R., McLeod, D., & Fox, M. D. (2019). Adverse childhood experiences and trust in the medical profession among young adults. *Journal of Health Care for the Poor and Underserved*, 30(1), 238–248.

Murray J. (2007). The cycle of punishment: Social exclusion of prisoners and their children. *Criminology & Criminal Justice*, 7(1), 55–81.

Murray, J., & Farrington, D. (2008). The effects of parental imprisonment on children. *Crime and Justice*, 37, 133–206.

Musa, A. A., & Ahmed, A. (2015). Criminal recidivism: A conceptual analysis of social exclusion. *Journal of Culture Society & Development*, 7, 28–34.

Nickel, C., Simek, M., Moleda, A., Muehlbacher, M., Buschmann, W., Fartacek, R., Bachler, E., Egger, C., Rother, W. K., Loew, T. H., & Nickel, M. K. (2006). Suicide attempts versus suicidal ideation in bulimic female adolescents. *Pediatrics International*, 48(4), 374–381.

Pfundmair, M. (2019). Ostracism promotes a terrorist mindset. *Behavioral Sciences of Terrorism and Political Aggression*, 11(2), 134–148.

Pozza, A., Albert, U., & Dèttore, D. (2020). Early maladaptive schemas as common and specific predictors of skin picking subtypes. *BMC Psychology*, 8(1), 1–11.

Price, J. (2015). *Prison and social death*. Rutgers University Press.

Reijntjes, A., Thomaes, S., Kamphuis, J. H., Bushman, B. J., De Castro, B. O., & Telch, M. J. (2011). Explaining the paradoxical rejection-aggression link: The mediating effects of hostile intent attributions, anger, and decreases in state self-esteem on peer rejection-induced aggression in youth. *Personality and Social Psychology Bulletin*, 37(7), 955–963.

Ren, D., Wesselmann, E. D., & Williams, K. D. (2018). Hurt people hurt people: Ostracism and aggression. *Current Opinion in Psychology*, 19, 34–38.

Riva, P., & Eck, J. (2016). The many faces of social exclusion. In P. Riva & J. Eck, *Social exclusion: Psychological approaches to understanding and reducing its impact* (pp. ix–xv). Springer International Publishing.

Riva, P., Montali, L., Wirth, J. H., Curioni, S., & Williams, K. D. (2017). Chronic social exclusion and evidence for the resignation stage: An empirical investigation. *Journal of Social and Personal Relationships*, 34(4), 541–564.

Riva, P., Williams, K. D., Torstrick, A. M., & Montali, L. (2014). Orders to shoot (a camera): Effects of ostracism on obedience. *The Journal of Social Psychology*, 154(3), 208–216.

Robertson, O., Christmann, K., Sharratt, K., Berman, A. H., Manby, M., Ayre, E., Foca, L., Asiminei, R., Philbrick, K., & Gavriluta, C. (2016). Children of prisoners: Their situation and role in long-term crime prevention. In H. Kury, S. Redo, & E. Shea (eds), *Women and children as victims and offenders: Background, prevention, reintegration* (pp. 203–232). Springer.

Social Exclusion Unit. (2002). *Reducing re-offending by ex-prisoners*. Social Exclusion Unit.

Sommer, F., Leuschner, V., & Scheithauer, H. (2014). Bullying, romantic rejection, and conflicts with teachers: The crucial role of social dynamics in the development of school shootings–A systematic review. *International Journal of Developmental Science*, 8(1–2), 3–24.

South, C. R., & Wood, J. (2006). Bullying in prisons: The importance of perceived social status, prisonization, and moral disengagement. *Aggressive Behavior: Official Journal of the International Society for Research on Aggression*, 32(5), 490–501.

Stearns, A. E., Swanson, R. & Etie, S. (2019) The walking dead? Assessing social death among long-term prisoners. *Corrections: Policy, Research and Practice*, 4(3), 153–168.

Steele, C., Kidd, D. C., & Castano, E. (2015). On social death: Ostracism and the accessibility of death thoughts. *Death Studies*, 39(1), 19–23.

Sykes, G. (1958). *The society of captives: A study of a maximum security prison*. Princeton University Press.

Twenge, J. M., Baumeister, R. F., DeWall, C. N., Ciarocco, N. J., & Bartels, J. M. (2007). Social exclusion decreases prosocial behavior. *Journal of Personality and Social Psychology*, 92(1), 56.

Twenge, J. M., Baumeister, R. F., Tice, D. M., & Stucke, T. S. (2001). If you can't join them, beat them: effects of social exclusion on aggressive behavior. *Journal of Personality and Social Psychology*, 81(6), 1058.

Twenge, J. M., Catanese, K. R., & Baumeister, R. F. (2003). Social exclusion and the deconstructed state: time perception, meaninglessness, lethargy, lack of emotion, and self-awareness. *Journal of Personality and Social Psychology*, 85(3), 409.

Van de Weijer, S. G., Smallbone, H. S., & Bouwman, V. (2018). Parental imprisonment and premature mortality in adulthood. *Journal of Developmental and Life-Course Criminology*, 4(2), 148–161.

Waytz, A., & Epley, N. (2012). Social connection enables dehumanization. *Journal of Experimental Social Psychology*, 48(1), 70–76.

Wesselmann, E. D., Ispas, D., Olson, M. D., Swerdlik, M. E., & Caudle, N. M. (2018). Does perceived ostracism contribute to mental health concerns among veterans who have been deployed? *PloS One*, 13(12), e0208438.

Wesselmann, E. D., & Williams, K. D. (2017). Social life and social death: Inclusion, ostracism, and rejection in groups. *Group Processes & Intergroup Relations*, 20(5), 693–706.

Williams, C. A., Doorley, J. D., & Esposito-Smythers, C. (2017). Interpersonal rejection sensitivity mediates the associations between peer victimization and two high-risk outcomes. *Clinical Child Psychology and Psychiatry*, 22(4), 649–663.

Williams, K. D. (2007). Ostracism: The kiss of social death. *Social and Personality Psychology Compass*, 1(1), 236–247.

Williams, K. D. (2009). Ostracism: A temporal need-threat model. In M. Zanna (ed.), *Advances in experimental social psychology* (pp. 279–314). Academic Press.

Williams, K. D., & Nida, S. A. (2011). Ostracism: Consequences and coping. *Current Directions in Psychological Science*, 20(2), 71–75.

Williams, K. & Nida, S. (2017). Introduction and overview. In K. Williams, & S. A. Nida (eds), *Ostracism, exclusion and rejection* (pp. 1–9). Routledge.

Williams, K. D., Cheung, C. K., & Choi, W. (2000). Cyberostracism: effects of being ignored over the Internet. *Journal of Personality and Social Psychology*, 79(5), 748.

Williams, K. D., Hales, A. H., & Michels, C. (2019). Social ostracism as a factor motivating interest in extreme groups. In S. C. Rudert, R. Greifeneder, & K. D. Williams (eds), *Current directions in ostracism, social exclusion, and rejection research* (pp. 18–31). Routledge.

Wyder, M., Ward, P., & De Leo, D. (2009). Separation as a suicide risk factor. *Journal of Affective Disorders*, 116(3), 208–213.

Young, J. E., Klosko, J. S., & Weishaar, M. E. (2003). *Schema therapy*. Guilford Press.

Zadro, L., Williams, K. D., & Richardson, R. (2004). How low can you go? Ostracism by a computer is sufficient to lower self-reported levels of belonging, control, self-esteem, and meaningful existence. *Journal of Experimental Social Psychology*, 40(4), 560–567.

10 Stereotyping and Prejudice

Derval Ambrose, Colin Campbell and Dennis Kaip

Introduction

Negative beliefs, attitudes, and behaviour, towards those who have come into contact with the criminal justice system (CJS) are widespread in society, and are frequently reflected in the media (Rosenberger & Callanan, 2012). Those who commit criminal offences violate one of the core values of civil society. The resulting beliefs, attitudes, and behaviour, in relation to such individuals may be underpinned by fear, but also by a need to be seen to uphold this core value by condemning those who have breached it in some way. However, such individuals can also trigger a voyeuristic curiosity, often fuelled by the fact that they are kept separate from society, and that most people will not have personal contact with them. This, in turn, can encourage the development of erroneous beliefs about individuals who have offended, and prejudicial attitudes and discriminatory behaviour based on these beliefs. These beliefs can also help us to manage those aggressive and antisocial aspects of ourselves that we would prefer not to acknowledge, or would like to get rid of, by locating them firmly in the person who has offended. By emphasising the difference between ourselves and those who commit criminal offences, there is a risk of both demonising this group of individuals, and distancing ourselves from them in a way that facilitates their dehumanisation.

Prejudice can be defined as "bias that devalues people because of their perceived membership of a social group" (Abrams, 2010, p. 8). In social psychological terms, prejudice is an attitude which is based on widely shared (usually negative) beliefs about, or internal representations of, a social group, known as stereotypes (Brown, 2010; Lippmann, 1922). Prejudice and stereotyping can result in an individual being treated differently due to their membership, or being perceived as having the characteristics, of a particular social group. This distancing, excluding, or denigrating behaviour *is* discrimination (Abrams, 2010).

This definition highlights two key characteristics of prejudice. Firstly, prejudice is best understood as a *process* which arises within an intergroup context, rather than as an attribute or characteristic of particular individuals

DOI: 10.4324/9781315560243-10

(Abrams & Christian, 2007). Secondly, it is mediated psychologically, in that it is based on an individual's interpretations of their social context, which, in turn, is influenced by their values, personality, and past experience (Abrams, 2010).

A significant predictor of both experiencing or perpetrating prejudice is whether people perceive themselves as belonging to a social category, or 'in-group', rather than as individuals. If they perceive themselves as belonging to a particular social category, they become invested in developing and maintaining a distinctive and positive in-group identity (Hogg & Abrams, 1988). Categorisation into superior and inferior social categories can serve the purpose of maintaining power and preventing others from accessing an equal share of resources. In doing so, this enables the group with the greatest power and access to resources to define the rules, frame the discourse, and identify and define those with less power. In this context, prejudice can be influenced by the perceived threat posed by another group, which may be real, symbolic, or economic (Stephan & Stephan, 2000), or by other factors such as the size of the group or ratio of in-group to out-group members (Abrams, 2010).

The development and expression of prejudice can be influenced by a wide range of factors, including values, or what is important to people in their lives. Prejudice can also be based on the perception that an out-group holds values that are contemptible or disgusting. Additionally, the extent to which an out-group appears not to appreciate a core value of the in-group can lead to the latter identifying the former as a legitimate target for prejudice (Abrams, 2010).

Once social categories have been defined, they can become invested with meaning that denotes power, status, and rights. While such stereotypical expectations can help to make life predictable, they can also be misapplied in a way that fuels prejudice. Applying implicit knowledge of social stereotypes based on individuals' perceived membership of social groups, can lead to certain inferences about them being drawn (Schneider, 2004). While these inferences may be inaccurate, they can help to both manage the perceived threat of out-groups, and inform prejudicial attitudes and discriminatory behaviour. This is often the case when the out-group is socially excluded (e.g. forensic clients), and there are limited opportunities for direct or sustained contact between the groups that might improve intergroup understanding, and, ultimately, undermine stereotypical beliefs. In the same way that social categorisation can be applied to others, it can also be applied to ourselves. Strong social identification with a particular social category can result in the embedding of an individual's identity largely within that category, and can be the basis of protracted prejudice, and intergroup conflict (Abrams, 2010).

Specific intergroup relations can have distinctive and enduring histories, and therefore unique problems which drive prejudice within these relationships. In this chapter, we will explore how membership of social categories related to gender, race, and class, influences attitudes and behaviour towards those who come into contact with the CJS, and how this can be both understood and addressed from a social psychological perspective.

Although we explore prejudice and stereotyping across three discrete areas, namely race and ethnicity, gender, and social class, the significant impact of intersectionality on adversity load is central to any consideration of prejudice. The concept of intersectionality originates from the work of feminist and critical theorists, and, in broad terms, describes the intercollective meaning and consequences of membership in various social groups (Crenshaw, 1989). Collins (2000) proposes that "intersectional paradigms suggest that certain ideas and/or practices surface repeatedly across multiple systems of oppression" (p. 47). Indeed, both systems of oppression and privilege can have a profound effect on an individual's risk of coming into contact with the CJS, as well as their experience of it. Throughout this chapter, we will explore how prejudice and stereotyping are relevant to forensic work, and encourage the reader to hold in mind the impact that multiple group memberships can have on experiences of oppression, and how this relates to forensic practice.

Race and Ethnicity

The first of the three discreet areas of prejudice and stereotyping that we will consider is that of race and ethnicity. Although stereotypes can be considered unavoidable, they can nonetheless lead to an erroneous generalisation that shapes one's understanding of an in-group and out-group, and which can, in turn, lead to prejudice (Chang & Kleiner, 2003; Fedor, 2014). Thus, in the context of race and ethnicity, considering the impact of stereotyping and prejudice is crucial.

Consideration of 'race' and 'ethnicity' is complicated. Not only do both terms share a difficult history, but they are often associated with negative connotations. Unfortunately, throughout history, and indeed across cultures, both terms have, for instance, been used to divide those who have privileges from those who do not, while also frequently promoting discrimination and prejudice (Smedley & Smedley, 2005). This history of separating people along racial and ethnic lines has led, and indeed still leads, to major intra- and/or inter-group conflicts, which can, in its most severe form, present as genocide (for example, on the genocide in Rwanda in 1994, see McDoom, 2013, 2014). Some argue that, based on the imbalances in power and privilege attributed to racial hegemony, race and ethnicity are distinct categories. Both terms have been used to contrast one group from another, and to highlight differences rather than similarities. However, added complexity comes from the fact that those terms are often used interchangeably (Cokley, 2007).

While the impact of racial discrimination is measurable, and the impact and effects of societal and institutional racism on the individual are well-documented, defining the term 'race' is rather difficult. Race is commonly understood as the separation of people into specific groups based on phenotypical traits, such as skin colour, facial features, or even hair type (Smedley & Smedley, 2005). However, it has been argued that race as a biological construct has no validity, and that dividing people merely by phenotypic traits is problematic as it constitutes historically-justified discrimination (Ford & Kelly, 2005; Naz et al., 2019; Walajahi et al., 2019). With no legitimate biological basis on which to support those distinctions, it has been proposed that the concept of race has rather been the result of social constructionism (Flanagin et al., 2021; Smedley & Smedley, 2005).

The term 'ethnicity' has been deemed equally difficult to conceptualise and define. The connections that bind people from one ethnic group to another are particularly difficult to detect and highlight. However, it is widely agreed that the term 'ethnicity', despite constantly evolving, stands separate from the term 'race' (Bartlett, 2001). People belonging to the same ethnicity are understood to have a putatively common heritage contingent upon a host of factors, such as cultural values, traditions, language, and norms (Lee, 2009). However, there is evidence that one key feature of ethnicity is the fact that it is often voluntarily defined by a group, and can be considered an achieved status, unlike race which is determined externally (Cornell & Hartmann, 2006; Lee, 2009).

Race and Ethnicity in Forensic Settings

In the UK, those from racialised minority communities are overrepresented in both secure forensic hospital and prison admissions, as well as across all stages within the CJS (Arya et al., 2021; Fernando et al., 1998; Lammy, 2017; Riggs Romaine & Kavanaugh, 2019). In particular, young Black men are considerably overrepresented within the youth justice system, as well as the wider CJS (Lammy, 2017). Beyond the UK, many other Western countries have an overrepresentation of their respective racialised minorities within their CJS. Indeed, both Australia and Canada have a disproportionate number of indigenous groups within their forensic settings (Carr, 2017; Pettit & Western, 2004).

Moreover, in comparison to their white peers, individuals from racialised minority groups are not only more likely to be diagnosed with serious mental illnesses, but are also more likely to be sent to prison rather than hospital (Perry et al., 2013; Pinals et al., 2004). Furthermore, despite the overrepresentation of the Black, Asian, and Minority Ethnic (BAME) community in prison and secure hospital settings, traditional therapeutic practices often fail to meet the needs of BAME groups (Naz et al., 2019).

The Lammy Review of 2017 in the UK investigated the outcomes and treatment of racialised minorities in the CJS (Lammy, 2017). It found that, in comparison to their white counterparts, BAME males were more likely to be arrested, 240 per cent more likely to go to prison, and more regularly categorised as high risk for comparable crimes. The review also established that there is a paucity of data on Gypsies, Roma, and Irish Traveller communities- groups who account for around 0.1 per cent of the general population but who make up approximately 5 per cent of the adult male prison population (Carr, 2017).

It is crucial to acknowledge that the terms BME (Black and Minority Ethnic) and BAME are highly debated, and that no clear consensus has yet been reached. Importantly, both are controversial terms for a myriad of reasons, including the fact that groups such as Gypsy, Roma, and Travellers of Irish Heritage, are not considered, and therefore experience further marginalisation (Condon et al., 2019). Thus, in this chapter the terms 'racially diverse' or 'racialised minorities' are used.

Positive Ethnic Identity

When considering the diversity of the forensic population, the practitioner needs to go beyond a basic understanding of what has shaped a client's experience of life. The practitioner should also focus on gaining an understanding of *how* a client has developed a sense of their ethnic identity, what support is needed to explore traumatic prejudicial experiences, and how any damage to the individual's sense of ethnic identity can be repaired. How can we support the discovery of one's culture and ethnicity? To what extent have constant experiences of stereotyping and prejudice impacted the individual's mental health, as well as their sense of positive ethnic identity? These and many other questions should be in the mind of the practitioner. Importantly, the culturally-competent practitioner must take into account the reality of trauma through racism, yet additionally clients have a range of positive identities and experiences relating to their ethnic background.

Phinney (1989) proposed a model of ethnic identity development (see also Huang & Stormshak, 2011) based on social identity theory (Tajfel 1981; Tajfel et al., 1979; chapter 3, this volume). Starting sometime in early childhood, one's sense of ethnic identity continues to develop throughout life (Umaña-Taylor et al., 2014). These are complex processes that entail constant internal debates about one's perception of oneself, and perceptions of others (Nagel, 1994). Furthermore, the development of ethnic identity requires an individual to examine their identity as a member of a certain group, which also necessitates cognitive development. Within Phinney's framework, the complex aspects of what constitutes ethnic identity are set out, in addition to the developmental stages that the individual might pass through to achieve a positive ethnic identity. These stages, presented in Figure 10.1, are not necessarily age-specific, and can present at any stage during adolescence. Additionally, progression from one stage to another does not happen automatically, and people can, in some instances, remain in one stage for their entire life. This model can help clinical practitioners to gain the understanding and cultural competence necessary to support their diverse patient groups (Hall & Jones, 2019).

During the 'unexamined ethnic identity' stage, an individual might not be particularly interested in their ethnic background. In turn, submitting, let alone conforming, to the overall attitude or perception of their surrounding culture may not be of interest to them at this stage (Phinney, 1993). The

Stage 1 - Unexamined Ethnic Identity

The extent of an individual's sense of belonging to their group

Stage 2 - Ethnic Identity Search (Moratorium)

Exploring activities that increase knowledge and experiences of one's ethnicity

Stage 3 - Ethnic Identity Achievement

Having a clear understanding of group membership and meaning of one's ethnicity

Figure 10.1 Phinney's stages of ethnic identity development
Source: Phinney (1989)

initiation of the second stage, 'moratorium', can be complex and based on a range of factors. In many instances, it can be the consequence of a challenging experience or crisis relating to their race or ethnicity, such as experiencing inequality, discrimination, and/or verbal and racial abuse (Phinney, 1993). This can be a catalyst for the process of exploring ethnic identity, which can then lead to the development of one's social consciousness and self-examination of one's ethnic group. This process of consciousness awakening and self-examination can either lead to the development of a sense of pride associated with belonging to that particular group, or to difficulties in coming to terms with the emerging identity (Phinney, 1993). This process can be very confusing and emotionally taxing for the individual, and has the potential to lead to mood changes, and increased anxiety and/or depression. The final stage, 'ethnic identity achievement', is when an individual attains a positive sense of ethnic identity and belonging to a specific group (Phinney, 1993). This acceptance is facilitated by the ability to identify both positive and less positive aspects of their identity, in tandem with recognising and accepting other ethnic identities and cultural backgrounds.

While ethnic identity is a crucial factor for developing a concept of self (Umaña-Taylor et al., 2014), it can be challenging for adolescents to develop it. One challenge might include being from an immigrant family, and having to navigate one's own cultural identity as well as the culture of the place of current residence (Berry & Sabatier, 2010; Cavdar et al., 2021). Many racialised minority families experience different sociocultural challenges, such as language barriers, acculturation, and discrimination, which inform the experiences and opinions of young people (Yasui & Dishion, 2007). Moreover, many young people are exposed to chronic stressors caused by prejudice and stereotypes based on their race and ethnicity.

To moderate the impact of these stressors and negative experiences, as well as to encourage growth, positive ethnic identity development is therefore deemed crucial for young people from racialised minority backgrounds (Williams et al., 2014). Yasui and Dishion (2007) propose that young people from racially diverse groups who have been able to establish a secure self-concept, can discover and acquire critical self-regulatory skills. Townsend and colleagues (2020) argue that a positive ethnic identity can operate as a moderating factor between trauma exposure and mental health diagnoses. Furthermore, positive ethnic identity is also an important protective factor that can reduce the negative effects of racial and ethnic discrimination (Yoo & Lee, 2008). Indeed, an extensive body of research supports the notion that adolescents and adults with a positive ethnic identity have greater self-esteem and self-efficacy, better developed coping skills, better behavioural outcomes, and, in some cases, better educational attainment (Huang & Stormshak, 2011; Rivas-Drake et al., 2014; Umaña-Taylor et al., 2014).

Evidence across different cultural contexts further illustrates the importance of a positive ethnic identity. For example, Bals and colleagues' study of the Sámi people, an indigenous people living in northern Scandinavia, found that achieving ethnic identity can be a stressful endeavour, possibly due to the fact that many adolescents in that region possess multiple ethnic identities (Bals et al., 2010). However, they found that native language competence, together with other forms of traditional knowledge, is a key element of ethnic identity, and is associated with positive psychosocial outcomes.

Further evidence can be found in Australia. A report on Australia's Northern Territory correctional services conducted by Hamburger et al. (2016), recommended, among other points, that incarcerated young Aboriginal People should have access to culture-specific practices, such as healing time and spirituality, before any offending-specific therapies. This was supported by the results of a later study on Aboriginal and Torres Strait Islander prisoners (Shepherd et al., 2018), in which a strong indigenous identity, concurrent with participation in cultural activities, was found to potentially reduce violent recidivism.

It is important too that white practitioners are able to reflect on their own identity formation, and consider what being 'white' has meant to them. Similarly, they should aim to explore what emotions might surface when confronted with navigating racial inequality, and reflect on how this might impact their own perceived identity. Indeed, the concept of white fragility (DiAngelo, 2015) is of relevance here as it relates to the possible defensive positions that might be assumed during the reflective process.

Box 10.1 Practitioner tip: Phinney's model

The establishment of a strong therapeutic relationship at the beginning of the therapeutic journey is paramount and should be a key objective for the practitioner when starting work with clients. By acknowledging and validating the clients' experiences as well as being prepared to learn their culture and ethnicity through their unique perspective, the practitioner is making a concerted effort towards positive rapport building.

This could be followed by the exploration of any potential traumatic experiences that the clients may have endured due to their ethnicity or cultural background. By utilising Phinney's (1989) model, the forensic practitioner, together with the client, could navigate those experiences, and continue the exploration process. This could be done through a range of activities, including reading and discussing the history of one's ethnic and cultural background, as well as elaborating on its impact on the individual, and, if possible, offering studies relating to their culture and ethnicity. Furthermore, the client could be supported and/or encouraged to attend or facilitate cultural events if appropriate. The aim is to move towards positive ethnic identity development.

Becoming a Culturally Competent Clinician

As the Western World transitions towards more diversified societies, it becomes evident that clinical practitioners must become culturally competent (Barber Rioja & Rosenfeld, 2018; Lee & Khawaja, 2013). Equipped with cultural competence, the skilled practitioner will be able to distinguish between whether a client's presentation is due to pathological or normative behaviours (Barber Rioja & Rosenfeld, 2018). This is particularly important in the context of forensic mental health where an inaccurate evaluation may not only cause legal ramifications but have deleterious consequences for both the individual and society (Barber Rioja & Rosenfeld, 2018; Kodjo, 2009). Thus, a culturally competent practitioner is able to tailor the treatment approach depending on their client's cultural or ethnic background and specific needs (Sue et al., 2009). Sue and colleagues (Sue et al., 1992, 2009) further posit that a culturally competent practitioner possesses specific multicultural capabilities that include awareness, knowledge, and skill. As alluded to above, this also includes the practitioners themselves who, within their therapeutic work, must reflect and engage with their own culture, attitudes, traditions, and biases (Lee & Khawaja, 2013).

Systemic Bias

It is important to acknowledge that the psychology discipline has a nefarious history of racism that negatively impacted on racialised minorities both at a client and practitioner level (Winston, 2020; Wood & Patel, 2017). Indeed, there is evidence that race and ethnicity can negatively influence those conducting forensic mental health assessments and reports in several ways (Riggs Romaine & Kavanaugh, 2019). For example, the efficacy of risk assessment tools across cultural groups is widely debated (Viljoen et al., 2018). Indeed, most studies on psychological assessments have been conducted on participants from Caucasian backgrounds, while diverse ethnic cohorts have been mostly disregarded (Snowden et al., 2010). Indeed, some literature has postulated that racial and ethnic differences demonstrably influence the predictive ability of various tools (Fass et al., 2008; Onifade et al., 2009).

Moreover, a study by Roberts et al. (2020) examined over 26,000 articles that were published in high-tier social, developmental, and cognitive journals between 1974 and 2018. Their study found that a paucity of psychological publications highlighted race as an important factor in human psychology, with an increasing presence in developmental psychology but a widely absent one in cognitive psychology. Furthermore, their findings indicated that the majority of publications were edited by white editors, and that numerous publications that were highlighting race were conducted by white authors, and typically based on participants who were mostly white. Thus, the validity of outcomes on a variety of psychological tests and research remains contested.

Racial and Ethnic Awareness in Clinical Practice

As previously established, the population in forensic settings is mostly male and comes from a culturally diverse background (Arya et al., 2021; Lammy, 2017; Fernando et al., 1998; Riggs Romaine & Kavanaugh, 2019), Yet the psychology and counselling profession consists mostly of white female practitioners (Baker & Nash, 2013; Chang & Berk, 2009; Odusanya et al., 2018). This homogeneity among practitioners, together with the clinician's potential lack of confidence in broaching the subject of race and ethnicity, could have deleterious consequences on the therapeutic process, and could risk the client's development if not addressed adequately (Chang & Yoon, 2011; Choi et al., 2015; Day-Vines et al., 2007). Consider the scenario between a clinical practitioner and their client shown in Figure 10.2.

The clinical practitioner can follow up in different ways. A common response from the practitioner could be: "Do you think you were potentially hypervigilant in those situations?" In contrast, a culturally competent practitioner could, together with their client, establish ways to examine and acknowledge prejudice and discrimination:

- "How does it feel to be treated so differently, and be subjected to such targeted extra scrutiny?"

Stereotyping and Prejudice 213

Clinical Practitioner:
White British female age 31.
Doctoral-level educated Clinical Psychologist.

Client:
Black British male age 52.
Grew up in the UK and left school age 15.

> I would like to start off by acknowledging that there are some differences between us. Would that be alright with you?

>> Not sure what you mean by that to be honest.

> For example, you identify yourself as a Black British male whilst I identify as a White British woman. I can imagine that we might have different experiences because of our backgrounds. I would like to create a space here where we can explore safely how those differences may have impacted on you and how they might affect our working relationship.

>> I appreciate that, but how could you ever possibly understand what I have experienced?

> That is true and I might get some things wrong, but I want you to feel safe enough to raise that with me. Not that I want to task you with my education rather it is important to me that you have the safety and freedom to talk to me about all the hurts that you have experienced in your lifetime.

>> I am not sure about this but I can try.

> Thank you. When do you remember first becoming aware of your race or ethnicity?

>> My whole life. It started when I was a little boy and it has not stopped since. You wouldn't understand. How could you anyways?

> Have you ever experienced a challenging or upsetting experience based on your race or ethnicity?

>> It can be normal everyday stuff. Like, the teacher gave me a harsher punishment than my peers for the same thing; being the only passenger on the train who gets his tickets checked; the bartender in the pub not wanting to serve me another because "I had enough" but serving the drunk white man; the police stopping me because "I match the description". Honestly, it happens so often.

Figure 10.2 A therapy dialogue

- "Those experiences of racism and discrimination could, very understandably, make you extra vigilant in certain situations and interactions. Do you think you might have developed a sophisticated internal alarm system to help you predict situations in which you may be unsafe?"
- "Those sound like very painful experiences. How does it impact you and what losses (such as access to opportunities, security, education, and sense of belonging) have you experienced because of it throughout your life?"
- "Many clients have talked of these losses as being a type of grief, does that resonate with you? I wonder if you have ever had the space to process some of that pain and grief?"

The acknowledgement of potential racial and ethnic differences between the patient and practitioner is crucial, as this might influence overall empathy and modes of communication (Bhui & Bhugra, 2004). Moreover, research has highlighted the importance of practitioners developing an understanding around barriers that racialised minority individuals can experience when trying to access mental health support (Iwamasa et al., 2002; Shin et al., 2016). Collaboratively working with clients is vital in determining how they identify with their ethnicity, and the impact this then has on their experiences.

Importantly, the historical context of race and ethnicity can make talking about it during therapy sessions particularly challenging for some practitioners, for fear of being insensitive (Cardemil & Battle, 2003; Naz et al., 2019). Compounding this further is the fact that clinical training often leaves practitioners underequipped to adequately deal with these scenarios. For example, a study by Hemmings and Evans (2018) examined 106 counselling practitioners, and found that, even though most clinicians have worked with individuals who had experienced race-based traumatisation, most participants did not have any adequate training to support those clients appropriately. Moreover, Iwamasa and colleagues suggested that a culturally competent clinician must, at the very least, have a basic understanding of the rich histories and complex heterogeneity of racialised minority cohorts (Iwamasa et al., 2002). Indeed, it is important to understand the client's lived experience of prejudice and discrimination, and the impact that this has had on them. Equally important is to examine one's own biases, as well as one's own ethnicity and cultural background. A structured method to introduce to a supervision session could be a web-based Implicit Association test (IAT) which has a variety of uses, including the measure of implicit biases, and bringing attention to potential prejudice (Hillard et al., 2013). Importantly, caution against irresponsible utilisation of the IAT and false interpretation of results is warranted (FitzGerald et al., 2019).

An additional process for meaningful supervision which enables the facilitation of such an important conversation has been proposed by Pendry (2012). This model asserts that, in the initial stage, the supervisor should create a safe space where their supervisee can, for example, have a conversation about their family origin, and engage in supported self-reflexivity. During the reflexivity stage, the supervisees are enabled to examine their cultural background, how their experiences were shaped by this, and what influence it has on their clinical practice (Hardy & Laszloffy, 1995). Importantly, it is the onus of the supervisor that their supervisees elaborate on questions about themselves that enables them to understand the influence of race and ethnicity in both supervision and clinical practice (Hardy, 2008, as cited in Pendry, 2012). Therefore, having a specific time set aside during supervision to learn and reflect on the impact of one's own cultural and ethnic background is crucial.

Furthermore, we would argue that cultural awareness and cultural competence should move towards the centre of future psychologist training and be a fundamental part of continued registration for psychological practitioners.

Gender

In the same way that prejudice and discrimination impact on individuals in the CJS due to their race and ethnicity, so too can it occur because of gender. Sexism, gender, and gender role stereotypes, have been an area of interest and extensive research for social psychologists for several decades (Broverman et al, 1972; Deaux & LaFrance, 1998; Fiske et al., 2007; Greenglass, 1982). In terms of relevance to forensic settings, we will focus on two main areas: firstly, the impact that both prejudice in the form of gender role stereotyping (or gender role norming) and concepts of masculinity can have on men and boys within the CJS; and, secondly, on how prejudice potentially plays out in the treatment of women and girls throughout the CJS. Within the scope of this chapter, we are examining gender from the broad position of males and females, largely as a reflection of how the CJS is structured. However, we recognise that gender is not a binary or dichotomous concept, and can be considered across multiple facets, including physiological, self-defined gender identity, legal gender, and gender expression (Lindqvist et al., 2021).

Men and Boys

While the gender of men and boys can grant them a certain level of privilege and power, it can also make them vulnerable to experiencing physical and psychological adversity. For example, in England and Wales in 2020, approximately 75 per cent of suicides were by men- an outcome which follows a consistent trend that has been evident since the mid-1990s (Office for National Statistics, 2020). Men are also disproportionately affected by illness in relation to alcohol misuse, with 61 per cent of alcohol-related hospital admissions in England, across 2017/18, being men (Office for National Statistics, 2019). Conversely, men are far less likely than women to access psychological therapies, illustrated by the finding that only 36 per cent of Improving Access to Psychological Therapies (IAPT) referrals in 2016/17 were for men (National Health Service, 2018).

The disproportional impact of gender starts young. Within the total number of children in local authority care in the UK, for example, boys almost always tend to be over-represented (Department for Education, 2019). Additionally, looked after children in the care system in England, are six times more likely to be convicted of a crime, or receive a caution than children outside of the care system (Department for Education, 2015). It is estimated that nearly a quarter of the adult prisoner population have been in care at some point during their childhoods (Williams et al., 2012). This is a stark statistic considering that approximately 1 per cent of the general population in the UK have been in the care system. Importantly, the experience of being in care can increase exposure to adverse and abusive experiences, such as child sexual abuse (CSA; Euser et al., 2013). It has been argued that CSA has often been understood as a feminist issue, and thus there is significant underreporting of the sexual abuse of boys (Briggs, 2007). In exploring the effectiveness of the Australian Protective Behaviours Programme, Briggs and Hawkins (1994a, 1994b, 1996) reported that 78.5 per cent of interviewed male victims believed that the abuse they were subjected to was 'normal' behaviour.

Set in the context of significant male over-representation in the CJS worldwide, this presents a concerning picture. Alarmingly, in September 2021, just over 95 per cent of the prison population in England and Wales were male (Ministry of Justice, 2021), and, in the USA, 89.5 per cent of the homicides between 1980 and 2008 were committed by men (Cooper & Smith, 2011).

Of course, males represent a diverse group, and adversity exposure, and experiences of prejudice and stereotyping, can be significantly affected by intersectionality, such as the impact of poverty, social class, race and ethnicity, migration status, sexual orientation, and gender identity- all of which can increase the adversity load. For example, as highlighted in the race and ethnicity section, despite people from racialised minority backgrounds making up 14 per cent of the UK population, they represent 25 per cent of the prison population (Lammy, 2017). This is even starker for young people, with 40 per cent of young people in custody being from a racialised minority background (ibid.).

Gender Role Strain

A useful social psychological model that can be used to consider the impact of prejudice and stereotyping on boys and men, is that of gender role strain. Pleck (1981) originally proposed the concept of sex role strain, which he later termed the Gender Role Strain Paradigm (GRSP; Pleck 1995), in part as a way to move past the more traditional gender role identity lens that had informed social psychological research on male gender from the 1930s onwards. Pleck (1995) defines gender role strain as a psychological state in which the demands of the gender role have adverse consequences for the individual. These negative consequences are both psychological and physical

(Pleck, 1981, 1995). He proposes that boys and men experience gender role strain when, for example, they deviate from gender role norms, fail to meet societal expected norms of masculinity, or experience internal discrepancies between real and ideal selves based on notions of gender norms. Furthermore, this strain is directly linked to internal and external expressions of prejudice, such as devaluing or restricting themselves, experiencing this from others, or imposing this on others (Pleck, 1995).

The GRSP interprets gender roles as socially constructed ideas that ultimately serve to maintain patriarchal social and economic power (Levant & Powell, 2017). This is firmly rooted in social psychological theories of stereotyping, including gender role stereotypes and norms, and the experience of prejudice based on adhering to, or failing to adhere to, these ideas (Brown, 2010). This interpretation in the GRSP represented a departure from the previous perspective on masculinity which was largely influenced by psychoanalytic theory, and adopted the position that humans have an innate psychological drive to conform to the gender identity that matches their biological sex. Pleck's (1995) move away from this approach was important as it allowed gender to be viewed through a lens that did not assume that it was the success or failure of meeting gender roles that determined psychological health. Rather, it allowed a view of gender that considered the distress caused to men by pressure resulting from systemic socialisation of men to have positions of power over women.

Pleck (1981, p.9) original description of the GRSP consisted of 10 propositions:

- Gender roles are operationally defined by gender role stereotypes and norms.
- Gender role norms are contradictory and inconsistent.
- The proportion of individuals who violate gender role norms is high.
- Violating gender role norms leads to social condemnation.
- Violating gender role norms leads to negative psychological consequences.
- Actual or imagined violation of gender role norms leads individuals to overconform to them.
- Violating gender role norms has more severe consequences for males than for females.
- Certain characteristics prescribed by gender role norms are psychologically dysfunctional.
- Each gender experiences gender role strain in its paid work and family roles.
- Historical change causes gender role strain.

Since this seminal piece of work, there has been four distinct clarifications to the original paradigm. Firstly, in relation to gender ideologies, Pleck (1995) has clarified that, although his original GRSP did not explicitly refer to this, the model places these ideologies at its core. Gender ideologies refer to societal beliefs about the importance of men and women conforming to dominant culturally defined roles of gender (Levant & Powell, 2017). Hence, individuals are shaped through social learning and societal influences throughout their lifespan. Pleck (1995) goes on to detail that there are identifiable standards and expectations that are attached to the concept of masculinity, and that these can often have negative connotations. David and Brannon's (1976) work on components of masculinity, while some time ago now, still holds true, and echoes much of the expectation we might see in clinical practice with men who have committed violence. They concluded that there were four components to the idea of being masculine- men should not be 'feminine', men should be concerned with striving to achieve, men should never show weakness, and men should seek adventure and risk, even if the latter results in violence. Secondly, clarification in relation to the social psychology of gender was largely to explicitly place the GRSP in a social psychological context. Thirdly, Pleck (1995) proposed three varieties of gender role strain, which are briefly summarised below:

- *Discrepancy Strain* – strain caused by an idea that one has not lived up to their internalised concept of ideal masculinity.
- *Dysfunction Strain* – strain is related to the impact that fulfilling the more traditionally masculine norms can have on men. That is, that several characteristics, traditionally viewed as desirable in men, can in fact impact men in significantly negative ways.
- *Trauma Strain* – strain impacting on men who have particularly severe experiences of gender role strain, such as men of colour, and survivors of childhood sexual abuse.

The final clarification to the GRSP was in relation to the social contexts of masculine gender role strain. This is, in essence, the recognition that context is important in terms of the expectations that men face, and, at times, different contexts may have different expectations of masculinity. This is important in terms of the experiences men may have, for example, in a prison environment where the environment is, arguably, hypermasculinised.

Implications for Forensic Practice

There has been a considerable amount of research that has explored the concepts covered above, and the impact they may have on male distress and aggression. Such literature indicates that perceived threats to masculinity are connected to increased anxiety (Vandello et al., 2008), lowered self-esteem (Ratliff & Oishi, 2013), increased victim blaming (Munsch & Willer, 2012), and increased general aggression (Talley & Bettencourt, 2008; Vandello et al., 2008). Reidy et al. (2015) developed a Masculine Gender Role Discrepancy Scale, and report links between high discrepancy stress and risky sexual behaviours, the perpetration of violence towards female intimate partners, and assault causing injury. Gebhard et al. (2019) suggest that shame may be the underlying mechanism that underpins aggressive responses to perceived threats to masculinity. They developed the Masculinity and Shame Questionnaire, and their research, with a sample of heterosexual men, concluded that there was a clear connection between threatened-masculinity shame-related responses, and self-reported tendencies to be physically aggressive. What this means for practitioners in forensic settings is important as it offers key contextual issues that need to be explored with men in the CJS in order to have any comprehensive understanding of the development and expression of aggression, distress, and interpersonal conflicts. Arguably, the prejudicial impact of gender role stereotypes and norms, resultant distress, and internalised shame, needs to be central to any formulation of risk. The following vignette is illustrative of the how this might present in forensic settings.

> Leroy grew up in an inner city area, raised by his mum and dad who had emigrated from the Caribbean when Leroy was an infant. Leroy started being bullied at school when he was 7 years old. He returned home from school one day injured and distressed after being punched in the face by another child. Leroy's father told him that boys don't cry and he needed to respond to bullies with increased violence. He added that if Leroy returned home again with an injury inflicted by another child, he would beat him twice as badly to help him become a man.
>
> By the time Leroy reached adolescence, he had been excluded from school, and had become part of a local gang in the estate where he lived. Leroy served several short sentences in YOI's for offences, including burglary and car theft. As an adult, Leroy continued to be in and out of prison for increasingly more serious offences. He has had difficulties maintaining employment and relationships. He has had some short-term relationships with women, and some sexual encounters with men, however he has kept the latter secret, and experiences deep-rooted shame about these encounters. Leroy is serving a 4-year sentence for assault, and is currently in the segregation unit following a vicious attack on a fellow prisoner who had made a sexual advance towards him.

A psychologist has been asked to complete a formulation of Leroy's risk of violence towards others. In order to place the impact of gender role stereotyping and experiences of prejudice at the centre of any narrative about Leroy's risk, the following should be considered:

- The impact that early life parenting and socialisation had on notions of masculinity, and the interplay this has with Leroy's attitudes towards violence.
- The distress and shame that Leroy has experienced when his emergent sexuality may have seemed to breach these notions of masculinity.

- The limits that were placed on emotional expression, in the context of gender role norms, from a young age, and how this has furthered internal distress, and increased feelings of being unsafe.
- The intersectional prejudices that Leroy likely experienced in relation to race, culture, immigration status, inner-city poverty, and sexuality.

As mentioned above, the concept of trauma strain implies a disproportionate impact on groups of men that may have a more severe experience of gender role norming; such as men of colour (Watkins et al., 2010), survivors of childhood sexual abuse (Lisak, 1995), and gay and bisexual men (Sánchez et al., 2010). Levant (1992) proposed that the restricted emotional expression encouraged in more extreme notions of masculinity may result in a mild to moderate alexithymia in men. This is illustrated in Leroy's case, and has obvious impacts on men's access to mental health support, including their low referral rates to mental health services as referred to above.

The impact of gender role strain is also crucial in our understandings of men who have been violent in intimate relationships. Reidy et al. (2014) assessed experiences of discrepancy stress, masculine gender role norms, and history of intimate partner violence (IPV) in a sample of 600 men. Results indicated that masculine discrepancy stress significantly predicted the sample's historic psychological, physical, and sexual IPV against women, independent of other masculinity variables. Additionally, Liang et al. (2017) explored gender dysfunction strain, and intervention programmes targeting men's intimate partner violence, substance misuse, and help-seeking behaviours. They concluded that, while gender dysfunction strain was considered in some programmes, there was clearly room to improve this, particularly given the small effect sizes, and high rates of recidivism seen in evaluations of such programmes. They recommend that one way to potentially enhance the effectiveness of these interventions is to take a gender transformative approach, whereby men are encouraged to learn about the impact of gender, transform gender roles, and seek more equitable gender relationships (ibid.). Furthermore, they conclude that, although critical in men's journeys to violence or distress, issues of intersectionality are poorly addressed in the interventions they reviewed.

The impact of the socialisation process for men on psychological distress, as well as the probability of seeking access to healthcare professionals to ameliorate this distress, is well-described. If men are socialised to not express, or be able to describe, their internal emotional world, the likely impact is increased isolation and distress. The impact that gender role stereotyping and prejudice has on the lives of men and boys needs to be more firmly addressed in clinical work with male forensic clients. Additionally, notions of masculinity and the distress that this can cause for men and boys should be central to harm reduction interventions. The GRSP presents a helpful model in which to consider these issues. The American Psychological Association (2018) published guidelines for psychological practice with boys and men, which offers a helpful steer for practitioners in the field, and include suggestions to recognise the social and cultural construction of gender identities, promote gender sensitive psychological services, seek systemic change, and recognise the impact of power, privilege, and sexism, on the development of boys and men (ibid.). Box 10.2 presents some tips that may be useful for practitioners in the field.

Box 10.2 Tips for practitioners

- If you have been socialised to not express vulnerability, then accessing this can feel threatening and ego dystonic.
- Building the therapeutic relationship and trust is paramount.
- Explore with male clients how aggression and violence has emerged through the lens of gender socialisation, stereotypes, and expectations.
- Recognise that violence reduction may not be ego syntonic for some men due to gender role strain, and that this conflict can hinder change.
- Explore discrepancy and shame.
- Recognise that certain groups of men will have had more traumatic experiences of masculinity socialisation and subsequent strains. Particular groups in clinical practice are often men who are survivors of childhood sexual abuse.

Women and Girls

Having considered the impact of stereotyping and prejudice on men and boys, what is its impact on women and girls? Worldwide, women and girls represent a small minority of the prison and wider forensic population (Walmsley, 2006). However, just as the overall imprisonment rates across most nations have increased over recent decades, so too have the number of incarcerated females. In the United States (US), incarceration rates have seen a fivefold increase since the 1970s, and have disproportionately impacted on women, with the number of women in prison in the US seeing a 14-fold increase in the same time period (Swavola et al., 2016). Similarly, in the UK, although women represent less than 5 per cent of the prison population, there has been a significant increase in the number of women sentenced to prison- with over 2200 more women in prison in England and Wales then there were 25 years ago (Prison Reform Trust, 2018).

Overwhelmingly, women and girls are imprisoned for non-violent crimes. Ministry of Justice (2019) figures report that 82 per cent of women, compared to 67 per cent of men, were imprisoned in England and Wales for non-violent crimes. Additionally, 48 per cent of women, versus just 22 per cent of men, committed their offence to support someone else's drug use (Light et al., 2013). It is estimated that 17,240 children are impacted by female imprisonment over a 12-month period (Wilks-Wiffen, 2011). Only 9 per cent of these children are looked after by their fathers while their mother is imprisoned (Corston, 2007). In conjunction with the separation from their children, women in prison in the UK also present with higher rates of mental health problems (Bartlett & Hollins, 2018), higher rates of self-harm (Ministry of Justice, 2018), significant histories of childhood emotional, sexual, and physical abuse (Williams et al., 2012), report higher rates of domestic violence victimisation (Ministry of Justice, 2014), and are often housed in prisons that are significant geographical distances from their families and support networks (Prison Reform Trust, 2016). Intersectionality is important in understanding how women are dealt with by the CJS, including the impact of race and ethnicity as covered above. For example, in contrast to white women, Asian women are 26 per cent more likely to be arrested, while Black women are 25 per cent more likely to receive a custodial sentence (Ministry of Justice, 2016).

In social psychology, the fundamental attribution error theory may offer a useful lens through which to consider how women are managed at all stages of their criminal justice journey. The fundamental attribution error was originally identified by Ross (1977), and refers to a tendency to attribute the cause of someone's behaviour to dispositional qualities, rather than to observable external factors. As such, when women or girls are seen to step outside of the expectations of their gender, and come in contact with the CJS, they can be seen as doubly deviant. That is, that women who offend have not only stepped outside of societal norms, but have also breached the norms of their gender (Heidensohn, 1968). As a result of this, woman may also be doubly punished (e.g. receive harsher sentences; Heidensohn & Silvestri, 2012). This may, in part, explain some of the higher rates of imprisonment for non-violent offending.

Implications for Forensic Practice

There are many different points in the CJS where gender stereotyping and prejudice can play a part, including arrest, sentencing, release, and probation supervision. Additionally, the impact of gender can be complicated, with evidence of both inter and intra gender biases. For example, Alderden and Ullman (2012) explored the impact of gender on detectives' arrest decision rates in sexual assault allegations in a US Midwestern police department. Interestingly, their findings suggest that, contrary to their hypothesis, female detectives were significantly less likely than their male counterparts to arrest suspects in sexual assault cases (Alderden & Ullman, 2012). This may, in part, be explained by women's experiences of growing up in patriarchal cultures where violence perpetrated by men against women can be perceived as an acceptable and expected norm.

The CJS was largely created *by* and *for* men, as they constitute the vast majority of people who are managed through the criminal justice process. This in turn impacts on the types of assessments and interventions that women have access to, with female forensic clients arguably being managed and treated through a system that often disregards gender theory. Similar to the white centric sampling issues we explored in race and ethnicity, most female clients are risk assessed using risk assessment tools that were designed for, and validated on, men. For example, the Level of Supervision Inventory-Revised (LSI-R; Andrews & Bonta, 2000) is widely used as an assessment of risk for women in the US. Similarly, in the UK, it is common that the HCR-20 (Douglas et al., 2013) is used to assess risk among women forensic clients detained in inpatient psychiatric units. Research into the use of these risk assessment tools for women, indicates that they lack accuracy. For example, Reisig et al. (2006) reported in their study that the LSI-R either over- or under-classified women forensic clients, with the result being that they were subjected to intensive interventions that were unnecessary, or received limited or no services. De Vogel et al. (2014) helpfully developed the Female Additional Manual (FAM), which can be applied when using the HCR-20 to address violence risk in women. This introduces items that are more evidenced-based as being related to female risk of violence (e.g. historic reviews of involvement in prostitution, and an assessment of self-esteem in relation to violence risk).

There have been interesting developments in terms of exploring potential gender specific pathways into crime (e.g. Brennan et al., 2012; Daly, 1992; Moffitt & Caspi, 2001). These explorations indicate that there are issues specific to the experiences of women (e.g. connections to trafficking and prostitution, and experiences of chronic domestic violence) that can contribute to their trajectory into criminal behaviour. In more recent years, there has been a concerted move towards providing gender-responsive approaches to female clients, which recognises these different pathways that women may proceed through before coming into contact with the CJS.

Box 10.3 Defining gender-responsiveness

Covington and Bloom (2000) define gender-responsiveness as:

> creating an environment through site selection, staff selection, program development, content, and material that reflects an understanding of the realities of women's lives and addresses the issues of the participants. Gender-responsive approaches are multidimensional and are based on theoretical perspectives that acknowledge women's pathways into the criminal justice system. These approaches address social (e.g. poverty, race, class, and gender inequality) and cultural factors, as well as therapeutic interventions. These interventions address issues such as abuse, violence, family relationships, substance abuse, and co-occurring disorders. They provide a strength-based approach to treatment and skill building. The emphasis is on self-efficacy.
>
> (Covington & Bloom, 2000, p. 11)

Bloom, Owen, and Covington (2005) suggest key guiding principles that should inform the management, supervision, and treatment, of women in the CJS. These include recognising that gender makes a difference, focusing on substance misuse, trauma, and mental health issues through comprehensive and culturally relevant services, and improving socioeconomic conditions for women. Gender-responsive services for women encourage a more systemic approach to recognising the role of gender stereotyping and prejudice.

Gender-responsive programmes have been piloted in the UK, the US, and Canada, and early research into their efficacy is promising. For example, Duwe and Clark (2015) evaluated the effectiveness of the Moving On programme delivered to females who had offended in Minnesota between 2001 and 2013. The Moving On programme is a gender-responsive cognitive-behavioural programme designed specifically for female forensic clients (Van Dieten, 2010; Van Dieten & MacKenna, 2001). They explored the efficacy of the programme across two distinct phases: firstly, when the programme was delivered with fidelity to the original design, and, secondly, when a shorter version of the programme was delivered. Findings suggest that, when delivered with fidelity, the Moving On programme lowered the risk of re-arrest and reconviction (Duwe & Clark, 2015).

Risk factors for both men and women can overlap significantly. For example, both groups can share experiences of childhood physical and sexual abuse, and/or poverty as detailed above. However, the key issue is that we cannot ignore the impact of gender on how people process and express the traumas that they have experienced. In particular, how central prejudice and stereotyping are to creating psychological distress. There are key differences, for example, in the types of violence that both groups can be exposed to, and their abilities to seek economic stability. Recognising prejudice and stereotyping on the basis of gender in service development and delivery seems key, as does the need for those entering our criminal justice systems to have a high-quality trauma-informed care and treatment which reflects issues that have been central to their gender, race, social class, culture, sexuality, and migrant status.

Class

In contrast to ethnicity and gender, the concept of class is relatively neglected in both social psychology, and when considering diversity and inclusion in public services, such as the CJS. This is despite evidence that individuals place high subjective importance on identities indicative of socioeconomic status (SES), at least as much as ethnicity and gender (Easterbrook et al., 2018, as cited in Manstead, 2018), and that it remains the best predictor of adult educational and occupational achievement (Jones, 2003).

While there is a perception that the class system has disappeared, the divisions between social classes, in terms of wealth and income inequality, are growing (Equality Trust, 2017; Office for National Statistics, 2014). Latent class analysis identifies seven social classes which are defined by financial circumstances, but also by social capital (i.e. the size of an individual's social network) and cultural capital (i.e. the extent of their engagement in different cultural activities; Savage et al., 2013). The same analyses identify the disappearance of the traditional, occupationally-defined working class, which now makes up only 14 per cent of the population, despite 60 per cent of the population still defining themselves as working class. In contrast, SES is defined in terms of economic position, educational attainment relative to others, and occupation. While individuals tend to think more readily in terms of SES, social psychology tends to use subjective social class and SES (Manstead, 2018). It is important to acknowledge that there is implied stereotyping and prejudice in the language that is used to describe class in social psychological research. This may be particularly the case as the term 'working class' becomes increasingly decoupled from its occupationally-based origins in relation to the means of production. Beliefs about, and attitudes towards, members of the working class are arguably most striking in US research, where the term 'lower class' tends to be used, rather than working class. In this chapter, we use these terms interchangeably and use the same terms as those used in primary sources where appropriate.

The Social Psychology of Class

Stephens et al. (2014) argue that social class gives rise to culture-specific selves and patterns of thinking, feeling, and acting. One type of self they identify is *expressive independence*, which is characteristic of those who grow up in affluent, middle-class contexts. In contrast to working class people, these individuals worry less about income or threats, and act in ways that reflect and reinforce their independence; "expressing their personal preferences, influencing their social contexts, standing out from others, and developing and exploring their own interests" (Stephens et al., 2014, p. 615). In contrast, *hard interdependence* is another type of self, which is characteristic of those growing up in low-income, working-class environments. Here, the self and behaviour are understood as interdependent with others and the social context, which are characterised by material constraints, less influence, choice, and control, and a need for resilience to cope with adversity. Contexts, such as home, school, and work, foster these self-conceptions. Standards for success, derived from expressive independence, create institutional barriers to social mobility in middle-class schools and workplaces. By promoting norms, which are more familiar to those who are middle-class, these institutions reproduce social inequalities while appearing to be meritocratic, and attribute the superior performance of the middle-class to ability and effort, rather than structural inequality (Stephens et al., 2014).

Distinctive patterns of behaviour based on differences in wealth, education, and occupation, can act as signals of social class, and, according to the subjective social rank argument (Kraus et al., 2011), can create cultural identities based on the subjective perception of social rank in relation to others. Such behavioural signals can also underpin prejudiced attitudes and discriminatory behaviour towards those from a lower social class. Kraus et al. (2012) argue that these differences in self-construal, based on subjective social rank, can shape social cognition. In other words, the way that we think about ourselves, or construct our sense of self, can influence the way that we process, store, and apply, information in social contexts, in order to explain and predict our own behaviour and that of others. For example, the way in which individuals from a lower-class background might think about the social environment is shaped by the perceived need to deal with external constraints and threats. A relative lack of security in terms of employment, housing, and personal safety, results in a heightened vigilance to threat, a lower sense of personal control over outcomes, and a preference for situational attributions. By comparison, individuals from upper class backgrounds might think about the social environment in terms of their own internal states, such as traits, emotions, and personal goals, rather than external forces outside of their control and influence (Kraus et al. 2012). As a result, they have a preference for dispositional attributions, which may, for example, encourage them to attribute poverty and unemployment to a lack of willpower or laziness, rather than to structural inequality or social injustice.

Consistent with the tendency of lower social class individuals to have more interdependent social relations, and to explain events in terms of external factors, such individuals typically score more highly on measures of empathy. Studies which have explored the relationship between social class and empathy have shown that individuals with lower educational attainment tend to score higher on a measure of empathic accuracy, and also on a measure of emotion recognition (Kraus et al., 2010). The possibility that being better at recognising the distress or need of others might influence the behaviour of lower social class individuals in situations where people are in distress or need, has also been explored. Lower social class individuals were more likely to trust, and allocate resources to, an anonymous other, and to support a charity (ibid.). Upper class individuals were more likely to be generous in a public context, have unethical decision-making tendencies, take valued goods from another, lie in a negotiation, and cheat to win a prize. The relationship between upper social class and these behaviours was mediated by favourable attitudes to greed and perceived benefit to self, and was increased in circumstances of high economic inequality, where there were more downward social class comparisons, more perceived threat, and a greater need to justify privilege.

Distinctive patterns in behaviour, cognition, and emotion, associated with subjective social class and SES can perpetuate structural disadvantage and inequality in society. One of the explanations for very low levels of working-class students at 'high-status' universities is that many do not apply because they do not anticipate that they will fit in in such institutions. That is, they perceive a mismatch between the identity conferred by their social background and the identity they associate with being a student at a university (Chowdry et al., 2013). Relatively disadvantaged students score lower on identity compatibility, which is associated with lower anticipated acceptance scores, and, in turn, predicts the type of university applied to, regardless of actual academic ability (Nieuwenhuis et al., 2019). The perceived mismatch between the interdependent norms of working-class individuals, and the independent norms of universities, can lead to greater discomfort, and poorer academic performance. Many perceive that the type of universities realistically available to them are second rate and may not result in the same employment opportunities open to those who graduate from 'high-status' universities (Hutchings & Archer, 2001). This may be compounded by a lack of the social and cultural capital needed to negotiate the application process for high-status jobs, and the impact of identity incompatibility on daily interactions with colleagues, which perpetuate disadvantage. The evidence appears consistent with these hypotheses, as individuals from poorer backgrounds are less likely to go to university than their more wealthy peers (Crawford et al., 2016). Those who do go to university are less likely to attend the ones of higher status, graduate, and achieve the highest classes of degree. The average earnings of graduates from poorer families are lower than that of graduates from more wealthy families, even when educational attainment, the university attended, and the subject studied, are controlled for (Crawford et al., 2016). In the context of considering the CJS, a study which looked at the educational backgrounds of the 'elite' professions across British society, found that senior judges in the UK were the profession with the highest independent school attendance, and the highest attendance at either Oxford or Cambridge Universities (Sutton Trust and Social Mobility Commission, 2019).

The power and resource imbalance inherent in the class system leads to the moral exclusion and dehumanisation of those in lower social classes, who are then treated with a lack of empathy and compassion. The dominant social response to the poor, and to poverty, is one of distancing, in which members of lower social classes are separated, excluded, devalued, discounted, and designated as 'other' (Lott, 2002). In social psychological terms, this is discrimination, which, together with stereotypes and prejudice, constitutes classism (Lott, 2012). Institutional distancing can be both deliberate and explicit or subtle and indirect, but is nonetheless seen within education, housing, healthcare, politics, public policy, and the CJS. There is disproportionate allocation of negative social value to the low status group, and barriers are put in place to obstruct full societal participation. In interpersonal terms, distancing is seen in the daily experience of being demeaned and discounted (Lott, 2002).

Class and the Criminal Justice System

Perhaps the most concrete example of distancing is imprisonment. Despite a relatively weak inverse link between socioeconomic status and criminality, those with low median income, and low levels of education are significantly overrepresented in the prison population (Social Exclusion Unit, 2002). Individuals who are perceived to be of lower social class are more likely to be perceived as a typical forensic client (Hoffmann, 1981), and as more likely to commit common, class-neutral offences (Smith et al., 2010). They are more likely to be convicted of a variety of crimes (Farrington, 1992; Flood-Page et al., 2000), and receive less attention if they are the victim of crime (Viano, 1992).

As discussed earlier, those of higher social class are more likely to make dispositional attributions in relation to lower class individuals (e.g. believing that they lack morals, and are to blame for the situation they find themselves in). Stereotypes of those from lower social classes are evident in the traits that college students endorse in relation to the poor. These include *angry, stupid, unpleasant, dirty, criminal, alcoholic, abusive,* and *violent* (Cozzarelli et al., 2001). Smith et al. (2013) argue that deeply conditioned disgust may underpin social distancing from those who are poor. In an experimental design, they found that the poor were more likely to be found guilty, and given a more severe punishment. Furthermore, crimes committed by the poor were more likely to be rated as disgusting, independent of their seriousness, particularly if the questionnaire item made reference to claiming welfare.

The anxiety, shame, and humiliation, experienced when the behaviour or appearance of those of lower social class violates middle class norms, is further evidence of interpersonal or institutional classism (Lott, 2012). However, complaint from those of lower social class is experienced as the individual being non-compliant or difficult, and, to preserve any sense of dignity, those of lower social class often simply accept the denial of resources. This is seen in the CJS, where those of lower social class lack the social and cultural capital to navigate the system, and the financial and social resources to secure the help of those who could help them. This is reflected in evidence that lawyers are reluctant to provide a service to the poor, and that their words and actions reflect negative, stereotyped beliefs about them (Merry, 1986). Distancing and classism are also evident in healthcare provision, including mental health services. For example, mental health professionals report feeling uncomfortable with low-income clients, and less able to empathise with them (Leeder, 1996). Lower social class clients are viewed as inarticulate, suspicious, resistant, apathetic, and passive. They are also more likely to receive therapy that is brief and drug-centred, and be treated by students or low-status professionals.

It is clear that professionals working in the CJS need to improve their awareness of the impact of classism on their work, and on those they work with. As an aspect of identity, it should be considered in the same way as gender and ethnicity in continuing professional development and training on diversity and inclusion. It is important that professionals are able to reflect on how the barriers created by classism maintain social inequality, and limit access to resources for those they work with from lower social classes. In doing so, they must consider their own role in maintaining and eliminating classist discrimination. This may include reflection on how middle-class professionals might benefit from maintaining the cycle of poverty, and how this might be maintained by models of working, which are predicated on the assumption that those of lower social class are different or inferior (Lott, 2002). For example, does a deficit model of offending behaviour or psychological well-being inevitably result in the behaviour of those trying to help reflecting cognitive and physical distance? Simple changes, such as sharing information on economic inequality, may help, as this has been shown to increase endorsement of government polices to reduce inequality, and strengthen the perception that economic success is due to structural factors rather than individual effort (McCall et al., 2017). Clinicians' roles as educators and supervisors also provide an important opportunity to improve awareness of classism in trainees and junior colleagues (Spence, 2012). The American Psychological Association (2008) has produced resources for inclusion of social class in psychology curricula, which can be used to improve recognition of social class as an independent variable that can have in an impact on behaviour and on the accessibility and effectiveness of interventions. Consideration of class can also be integrated into supervision. Smith (2009) has suggested five interrelated areas that supervisors and trainees can explore in order to better consider issues of class and social inequality within their continuing professional development (see Box 10.4).

Box 10.4 Enhancing competence in the context of poverty

Five areas for action:

1. Recommend supplemental readings on social class issues.
2. Help supervisees consider their class privilege (as you consider your own).
3. Support supervisees in processing a 'Realm of Trauma'.
4. Implement a social justice framework within supervision.
5. Support flexible approaches to treatment.

Conclusion

It is clear that both prejudice and stereotyping play important roles in how difference can impact on the development of risk factors for offending, as well as negative experiences of forensic systems. Additionally, those who come into contact with the CJS represent a group that are already significantly maligned and marginalised. For practitioners, the impact of difference, such as race and ethnicity, gender, and class, should be central to assessment and treatment. This is further highlighted by the impact that multiple diversities and differences can have on an individual's experiences and journey through forensic systems. Consider, for example, how intersectionality can impact transgender clients in detained settings. Maycock (2020) explored the experiences of a small sample of transgender prisoners in Scotland, utilising Sykes (1958) *pains of imprisonment* lens. He reported that the sample referred to a perception that staff undermined their pursuit of authenticity, the difficulty of being located in an environment that matched sex assigned at birth, and how this affected their sense of self, difficulties in accessing necessary healthcare appointments, and experiencing transphobia in prisons as more intense to that in community settings (Maycock, 2020).

The potential psychological injury that frequent experiences of prejudice and stereotyping can have on an individual has important implications for clinical practice. For example, how chronic experiences of racism can result in significant levels of internalised shame. Johnson (2020) explored this area with a sample of African American college students, and reported that past year and lifetime experiences of racism were positively associated with internalised shame. Practitioners should provide a safe space for clients in which they can explore their experiences of prejudice, as well as an opportunity to develop positive ethnic identities. Similarly, gender and class, and related stereotyping and prejudice, needs to be considered for forensic clients. Men, racialised minorities, and those from lower socio-economic backgrounds, are over-represented in many prison populations. Arguably, rehabilitation needs to not just acknowledge this, but also reflect this in all areas of clinical practice. In more recent years, the principles of Trauma-Informed Care have become more commonplace in some forensic settings (Harris & Fallot, 2001). This perhaps represents a useful framework that can be used to begin addressing the cumulative internalised experiences of prejudice that the forensic client group have had.

References

Abrams, D. (2010). Processes of prejudice: Theory, evidence and intervention. Retrieved from www.equalityhumanrights.com/sites/default/files/research-report-56-processes-of-prejudice-theory-evidence-and-intervention.pdf

Abrams, D., & Christian, J. N. (2007). A relational analysis of social exclusion. In D. Abrams, J. N.Christian, & D. Gordon (eds), *Multidisciplinary handbook of social exclusion research* (pp. 211–232). Wiley-Blackwell.

Alderden, M. A., & Ullman, S. E. (2012). Gender difference or indifference? Detective decision making in sexual assault cases. *Journal of Interpersonal Violence*, 27(1), 3–22.

American Psychological Association. (2008). Report of the APA Task Force on Resources for the Inclusion of Social Class in Psychology Curricula. Retrieved from www.apa.org/pi/ses/resources/publications/social-class-curricula.

American Psychological Association. (2018). APA guidelines for psychological practice with boys and men. Retrieved from www.apa.org/about/policy/psychological-practice-boys-men-guidelines.pdf.

Andrews, D. A., & Bonta, J. L. (2000). *The level of service inventory-revised*. Multi-Health Systems.

Arya, D., Connolly, C., & Yeoman, B. (2021). Black and minority ethnic groups and forensic mental health. *BJPsych Open*, 7(S1), S123–S123.

Bals, M., Turi, A. L., Skre, I., & Kvernmo, S. (2010). Internalization symptoms, perceived discrimination, and ethnic identity in indigenous Sami and non-Sami youth in Arctic Norway. *Ethnicity & Health*, 15(2), 165–179.

Baker, M., & Nash, J. (2013). Women Entering Clinical Psychology: Q-Sort Narratives of Career Attraction of Female Clinical Psychology Trainees in the UK. *Clinical Psychology & Psychotherapy*, 20(3), 246–253.

Barber Rioja, V., & Rosenfeld, B. (2018). Addressing linguistic and cultural differences in the forensic interview. *International Journal of Forensic Mental Health*, 17(4), 377–386.

Bartlett, R. (2001). Medieval and modern concepts of race and ethnicity. *Journal of Medieval and Early Modern Studies*, 31(1), 39–56.

Bartlett. A., & Hollins, S. (2018). Challenges and mental health needs of women in prison. *The British Journal of Psychiatry*, 212(3), 134–136.

Berry, J. W., & Sabatier, C. (2010). Acculturation, discrimination, and adaptation among second generation immigrant youth in Montreal and Paris. *International Journal of Intercultural Relations*, 34(3), 191–207.

Bhui, K., & Bhugra, D. (2004). Communication with patients from other cultures: the place of explanatory models. *Advances in Psychiatric Treatment*, 10(6), 474–478.

Bloom, B., Owen, B., & Covington, S. (2005). *Gender-responsive strategies for women offenders: a summary of research practice, and guiding principles for women offenders*. US Department of Justice, National Institute of Corrections.

Brennan, T., Breitenbach, M., Dieterich, W., Salisbury, E. J., & van Voorhis, P. (2012). Women's pathways to serious and habitual crime: A person-lefted analysis incorporating gender responsive factors. *Criminal Justice and Behavior*, 39(11), 1481–1508.

Briggs, F., & Hawkins, R. (1994a). Follow up data on the effectiveness of New Zealand's national school based child protection program. *Child Abuse and Neglect, The International Journal*, 18(8), 635–643.

Briggs, F., & Hawkins, R. (1994b). Choosing between child protection programmes. *Child Abuse Review*, 3(4), 272–284.

Briggs, F., & Hawkins, R. M. F. (1996). A comparison of the childhood experiences of convicted male child molesters and men who were sexually abused in childhood and claimed to be non-offenders. *Child Abuse and Neglect, The International Journal*, 20(3), 221–234.

Briggs, F. (2007). The challenge of protecting boys from sexual abuse. Retrieved from https://apo.org.au/sites/default/files/resource-files/2008-01/apo-nid3873.pdf.

Broverman, I. K., Vogel, S. R., Broverman, D. M., Clarkson, F. E., & RosenKrantz, P. S. (1972). Sex-role stereotypes and clinical judgements of mental health: A current appraisal. *Journal of Social Issues*, 28, 59–78.

Brown, R. (2010). *Prejudice: Its social psychology*. Wiley-Blackwell.

Cardemil, E. V., & Battle, C. L. (2003). Guess who's coming to therapy? Getting comfortable with conversations about race and ethnicity in psychotherapy. *Professional Psychology: Research and Practice*, 34(3), 278–286.

Carr, N. (2017). The Lammy Review and race and bias in the criminal justice system. *Probation Journal*, 64(4), 333–336.

Cavdar, D., McKeown, S., & Rose, J. (2021). Mental health outcomes of ethnic identity and acculturation among British-born children of immigrants from Turkey. *New Directions for Child and Adolescent Development*, 176, 141–161.

Chang, D. F., & Berk, A. (2009). Making cross-racial therapy work: A phenomenological study of clients' experiences of cross-racial therapy. *Journal of Counseling Psychology*, 56(4), 521–536.

Chang, S. H., & Kleiner, B. H. (2003). Common racial stereotypes. *Equal Opportunities International*, 22(3), 1–9.

Chang, D. F., & Yoon, P. (2011). Ethnic minority clients' perceptions of the significance of race in cross-racial therapy relationships. *Psychotherapy Research*, 21(5), 567–582.

Choi, G., Mallinckrodt, B., & Richardson, J. D. (2015). Effects of international student counselors' broaching statements about cultural and language differences on participants' perceptions of the counselors. *Journal of Multicultural Counseling and Development*, 43(1), 25–37.

Chowdry, H., Crawford, C., Dearden, L., Goodman, A., & Vignoles, A. (2013). Widening participation in higher education: Analysis using linked administrative data. *Journal of the Royal Statistical Society: Series A*, 176, 431–457.

Cokley, K. (2007). Critical issues in the measurement of ethnic and racial identity: A referendum on the state of the field. *Journal of Counseling Psychology*, 54(3), 224–234.

Collins, P. H. (2000). Gender, black feminism, and black political economy. *The Annals of the American Academy of Political and Social Science* 568(1), 41–53.

Condon, L., Bedford, H., Ireland, L., Kerr, S., Mytton, J., Richardson, Z., & Jackson, C. (2019). Engaging Gypsy, Roma, and traveller communities in research: maximizing opportunities and overcoming challenges. *Qualitative Health Research*, 29(9), 1324–1333.

Cooper, A., & Smith, E. L. (2011). Homicide trends in the United States, 1980–2008 annual rates for 2009 and 2010. Retrieved from https://bjs.ojp.gov/content/pub/pdf/htus8008.pdf.

Cornell, S., & Hartmann, D. (2006). *Ethnicity and race: Making identities in a changing world*. Sage Publications.

Corston, J. (2007). *The Corston report: A report by Baroness Jean Corston of a review of women with particular vulnerabilities in the criminal justice system*. The Home Office.

Covington, S., & Bloom, B. (2000). Gendered justice: Programming for women in correctional settings. Retrieved from www.stephaniecovington.com/site/assets/files/1542/11.pdf

Cozzarelli, C., Wilkinson, A. V., & Tagler, M. J. (2001). Attitudes toward the poor and attributions for poverty. *Journal of Social Issues*, 57, 207–228.

Crawford, C., Gregg, P., Macmillan, L., Vignoles, A. & Wyness, G. (2016). Higher education, career opportunities, and intergenerational inequality. *Oxford Review of Economic Policy*, 32(4), 553–575.

Crenshaw, K. (1989). Demarginalizing the intersection of race and sex: A Black feminist critique of antidiscrimination doctrine, feminist theory, and antiracist politics. *University of Chicago Legal Forum*, 1989(1), 139–167.

Daly, K. (1992). Women's pathways to felony court: Feminist theories of lawbreaking and problems of representation. *Review of Law and Women's Studies*, 2, 11–52.

David, D., & Brannon, R. (1976). The male sex role: Our culture's blueprint of manhood, and what it's done for us lately. In D. David & R. Brannon (eds), *The forty-nine percent majority: The male sex role* (pp. 1–48). Addison-Wesley.

Day-Vines, N. L., Wood, S. M., Grothaus, T., Craigen, L., Holman, A., Dotson-Blake, K., & Douglass, M. J. (2007). Broaching the subjects of race, ethnicity, and culture during the counseling process. *Journal of Counseling & Development*, 85(4), 401–409.

Deaux, K., & LaFrance, M. (1998). Gender. In D. T. Gilbert, S. T. Fiske, & G. Lindzey (eds), *The handbook of social psychology* (4th edition, Vol. 1, pp. 788–827). McGraw-Hill.

De Vogel, V., de Vries Robbé, M., van Kalmthout, W., & Place, C. (2014). *Female Additional Manual (FAM): Additional guidelines to the HCR-20V3 for assessing risk for violence in women. English version.* Van der Hoeven Kliniek.

Department for Education. (2015). Outcomes for children looked after by local authorities, as at 31 March 2014. Retrieved from https://assets.publishing.service.gov.uk/government/uploads/system/uploads/attachment_data/file/384781/Outcomes_SFR49_2014_Text.pdf

Department for Education. (2019). Children looked after in England (including adoption) year ending 31 March 2019. Retrieved from https://assets.publishing.service.gov.uk/government/uploads/system/uploads/attachment_data/file/850306/Children_looked_after_in_England_2019_Text.pdf

DiAngelo, R. (2015). White fragility: Why it's so hard to talk to White people about racism. Retrieved from http://goodmenproject.com/featured-content/white-fragility-why-its-so-hard-to-talk-to-white-people-about-racism-twlm/.

Douglas, K. S., Hart, S. D., Webster, C. D., & Belfrage, H. (2013). *HCR-20V3: Assessing risk of violence – User guide.* Mental Health, Law, and Policy Institute, Simon Fraser University.

Duwe, G., & Clark, A. C. (2015). Importance of program integrity: Outcome evaluation of a gender-responsive, cognitive-behavioral program for female offenders, *Criminology & Public Policy*, 14(2), 301–328.

Equality Trust. (2017). How has inequality changed? Retrieved from https://equalitytrust.org.uk/how-has-inequality-changed.

Euser, S., Alink, L. R., Tharner, A., van IJzendoorn, M. H., & Bakermans-Kranenburg, M. J. (2013). The prevalence of child sexual abuse in out-of-home care: A comparison between abuse in residential and in foster care. *Child Maltreatment*, 18(4), 221–231.

Farrington, D. P. (1992). Explaining the beginning, progress and ending of antisocial behaviour from birth to adulthood. In J. McCord (ed.), *Facts, frameworks and forecasts: Advances in criminological theory* (pp. 253–286). Transaction Publishers.

Fass, T. L., Heilbrun, K., DeMatteo, D., & Fretz, R. (2008). The LSI-R and the COMPAS: Validation data on two risk-needs tools. *Criminal Justice and Behavior*, 35(9), 1095–1108.

Fedor, C. G. (2014). Stereotypes and prejudice in the perception of the 'other'. *Procedia-Social and Behavioral Sciences*, 149, 321–326.

Fernando, S., Ndegwa, D., & Wilson, M. (1998). *Forensic psychiatry, race and culture.* Psychology Press.

Fiske, S. T., Cuddy, A., & Glick, P. (2007). Universal dimensions of social perception: Warmth and competence. *Trends in Cognitive Science*, 11, 77–83.

FitzGerald, C., Martin, A., Berner, D., & Hurst, S. (2019). Interventions designed to reduce implicit prejudices and implicit stereotypes in real world contexts: a systematic review. *BMC Psychology*, 7(1), 1–12.

Flanagin, A., Frey, T., Christiansen, S. L., & AMA Manual of Style Committee. (2021). Updated guidance on the reporting of race and ethnicity in medical and science journals. *JAMA*, 326(7), 621–627.

Flood-Page, C., Campbell, S., Harrington, V., & Miller, J. (2000). *Youth crime: Findings from the 1998/99 youth lifestyles survey.* Home Office.

Ford, M. E., & Kelly, P. A. (2005). Conceptualizing and categorizing race and ethnicity in health services research. *Health Services Research*, 40, 1658–1675.

Gebhard, K. T., Cattaneo, L. B., Tangney, J. P., Hargrove, S., & Shor, R. (2019). Threatened-Masculinity shame-related responses among straight men: Measurement and relationship to aggression. *Psychology of Men & Masculinity*, 20, 429–444.

Greenglass, E. R. (1982). *A world of difference: Gender roles in perspective*. Wiley.

Hall, E. L., & Jones, N. P. (2019). A deeper analysis of culturally competent practice: Delving beneath white privilege. *Journal of Ethnic & Cultural Diversity in Social Work*, 28(3), 282–296.

Hamburger, K., Ferris, A., Hocken, J., Downes, L., Ellis-Smith, T., & McAllister, N. (2016). *A safer northern territory through correctional interventions: Report of the review of the northern territory department of correctional services*. BDO/Knowledge Consulting. Retrieved from https://apo.org.au/sites/default/files/resource-files/2016-11/apo-nid70848.pdf.

Hardy, K. V., & Laszloffy, T. A. (1995). The cultural genogram: Key to training culturally competent family therapists. *Journal of Marital and Family Therapy*, 21(3), 227–237.

Harris, M., & Fallot, R. D. (2001). Envisioning a trauma-informed service system: a vital paradigm shift. *New Directions for Mental Health Services*, 2001(89), 3–22.

Heidensohn, F. (1968). The deviance of women: A critique and an enquiry. *The British Journal of Sociology*, 19(2), 160–175.

Heidensohn, F., & Silvestri, M. (2012). Gender and crime. In M. Maguire, R. Morgan, & R. Reiner (eds), *The Oxford handbook of criminology* (5th edition, pp. 336–369). Oxford University Press.

Hemmings, C., & Evans, A. M. (2018). Identifying and treating race-based trauma in counseling. *Journal of Multicultural Counseling and Development*, 46(1), 20–39.

Hillard, A. L., Ryan, C. S., & Gervais, S. J. (2013). Reactions to the implicit association test as an educational tool: A mixed methods study. *Social Psychology of Education*, 16(3), 495–516.

Hoffmann, E. (1981). Social class correlates of perceived offender typicality. *Psychological Reports*, 49, 347–350.

Hogg, M.A., & Abrams, D. (1988). *Social identifications: A social psychology of intergroup relations and group processes*. Routledge.

Huang, C. Y., & Stormshak, E. A. (2011). A longitudinal examination of early adolescence ethnic identity trajectories. *Cultural Diversity & Ethnic Minority Psychology*, 17(3), 261–270.

Hutchings, M., & Archer, L. (2001). 'Higher than Einstein': Constructions of going to university among working-class non-participants. *Research Papers in Education*, 16, 69–91.

Iwamasa, G. Y., Sorocco, K. H., & Koonce, D. A. (2002). Ethnicity and clinical psychology: A content analysis of the literature. *Clinical Psychology Review*, 22(6), 931–944.

Iwamoto, D. K., & Liu, W. M. (2010). The impact of racial identity, ethnic identity, Asian values, and race-related stress on Asian Americans and Asian international college students' psychological well-being. *Journal of Counselling Psychology*, 57(1), 79–91.

Johnson, A. J. (2020). Examining associations between racism, internalized shame, and self-esteem among African Americans, *Cogent Psychology*, 7(1), article 1757857.

Jones, S. J. (2003). Complex subjectivities: Class, ethnicity and race in women's narratives of upward mobility. *Journal of Social Issues*, 59, 803–820.

Kodjo, C. (2009). Cultural competence in clinician communication. *Pediatrics in Review*, 30(2), 57–64.

Kraus, M. W., Cote, S., & Keltner, D. (2010). Social class, contextualism, and empathic accuracy. *Psychological Science*, 21, 1716–1723.

Kraus, M. W., Piff, P. K., & Keltner, D. (2011). Social class as culture: The convergence of resources and rank in the social realm. *Current Directions in Psychological Science*, 20, 246–250.

Kraus, M. W., Piff, P. K., Mendoza-Denton, R., Rheinschmidt, M. L., & Keltner, D. (2012). Social class, solipsism, and contextualism: How the rich are different from the poor. *Psychological Review*, 119, 546–572.

Lammy, D. (2017). *The Lammy review: An independent review into the treatment of, and outcomes for, Black, Asian and Minority Ethnic individuals in the criminal justice system.* Department of Justice.

Lee, C. (2009). "Race" and "ethnicity" in biomedical research: how do scientists construct and explain differences in health? *Social Science & Medicine*, 68(6), 1183–1190.

Lee, A., & Khawaja, N. G. (2013). Multicultural training experiences as predictors of psychology students' cultural competence. *Australian Psychologist*, 48(3), 209–216.

Leeder, E. (1996). Speaking rich people's words: Implications of a feminist class analysis and psychotherapy. In M. Hill & E. Rothblum (eds), *Classism and feminist therapy: Counting costs* (pp. 45–57). Haworth.

Levant, R. F. (1992). Toward the reconstruction of masculinity. *Journal of Family Psychology*, 5, 379–402.

Levant, R. F., & Powell, W. A. (2017). The gender role strain paradigm. In R. F. Levant & Y. J. Wong (eds), *The psychology of men and masculinities* (pp. 15–43). American Psychological Association.

Liang, C. T. H., Molenaar, C., Hermann, C., & Rivera, L. A. (2017). Dysfunction strain and intervention programs aimed at men's violence, substance use, and help-seeking behavior. In R. F. Levant & Y. J. Wong (eds), *The psychology of men and masculinities* (pp. 347–377). American Psychological Association.

Light, M., Grant, E., & Hopkins, K. (2013). *Gender differences in substance misuse and mental health amongst prisoners: Results from the Surveying Prisoner Crime Reduction (SPCR).* Ministry of Justice.

Lindqvist, A., Sendén, M. G., & Renström, E. A. (2021). What is gender, anyway: a review of the options for operationalising gender, *Psychology & Sexuality*, 12(4), 332–344.

Lippmann, W. (1922). *Public opinion.* Harcourt-Brace.

Lisak, D. (1995). *Integrating gender analysis in psychotherapy with male survivors of abuse.* Paper presentation, the Convention of the American Psychological Association, New York.

Lott, B. (2002). Cognitive and behavioral distancing from the poor. *American Psychologist*, 57(2), 100–110.

Lott, B. (2012). The social psychology of class and classism. *American Psychologist*, 67(8), 650–658.

Manstead, A. S. R. (2018). The psychology of social class: How socioeconomic status impacts thought, feelings and behaviour. *British Journal of Social Psychology*, 57, 267–291.

Maycock, M. (2020). The transgender pain of imprisonment. *European Journal of Criminology*, online ahead of print.

McCall, L., Burk, D., Laperriere, M., & Richeson, J. A. (2017). Exposure to rising inequality shapes Americans' opportunity beliefs and policy support. *Proceedings of the National Academy of Sciences*, 114, 9593–9598.

McDoom, O. S. (2013). Who killed in Rwanda's genocide? Micro-space, social influence and individual participation in intergroup violence. *Journal of Peace Research*, 50(4), 453–467.

McDoom, O. S. (2014). Predicting violence within genocide: A model of elite competition and ethnic segregation from Rwanda. *Political Geography*, 42, 34–45.

Merry, S. E. (1986). Everyday understanding of the law in working-class America. *American Ethnologist*, 13, 253–270.

Ministry of Justice. (2014). *Thinking differently about female offenders, Transforming Rehabilitation, Guidance document*. MOH/NOMS.

Ministry of Justice. (2016). Black, Asian and Minority Ethnic disproportionality in the Criminal Justice System in England and Wales. Retrieved from https://assets.publishing.service.gov.uk/government/uploads/system/uploads/attachment_data/file/639261/bame-disproportionality-in-the-cjs.pdf

Ministry of Justice. (2018). Statistics on Women and the Criminal Justice System 2017: A Ministry of Justice publication under Section 95 of the Criminal Act 1991. Retrieved from https://assets.publishing.service.gov.uk/government/uploads/system/uploads/attachment_data/file/759770/women-criminal-justice-system-2017.pdf.

Ministry of Justice. (2019). Offender management statistics quarterly: October to December 2018. Retrieved from https://assets.publishing.service.gov.uk/government/uploads/system/uploads/attachment_data/file/805271/offender-management-statistics-quarterly-q4-2018.pdf.

Ministry of Justice. (2021). Offender management statistics quarterly. Retrieved from www.gov.uk/government/collections/offender-management-statistics-quarterly.

Moffitt, T., & Caspi, A. (2001). Childhood predictors differentiate life-course persistent and adolescence-limited anti-social pathways among males and females, *Development and Psychopathology*, 13(2), 355–375.

Munsch, C. L., & Willer, R. (2012). The role of gender identity threat in perceptions of date rape and sexual coercion. *Violence Against Women*, 18, 1125–1146.

Nagel, J. (1994). Constructing ethnicity: Creating and recreating ethnic identity and culture. *Social problems*, 41(1), 152–176.

National Health Service. (2018). Psychological therapies: Annual report on the use of IAPT services England, further analyses on 2016–17. Retrieved from https://digital.nhs.uk/data-and-information/publications/statistical/psychological-therapies-annual-reports-on-the-use-of-iapt-services/annual-report-2016-17-further-analyses.

Naz, S., Gregory, R., & Bahu, M. (2019). Addressing issues of race, ethnicity and culture in CBT to support therapists and service managers to deliver culturally competent therapy and reduce inequalities in mental health provision for BAME service users. *The Cognitive Behaviour Therapist*, 12, E22.

Nieuwenhuis, M., Easterbrook, M., & Manstead, A. S. R. (2019). Accounting for unequal access to higher education: The role of social identity factors. *Group Processes & Intergroup Relations*, 22(3), 371–389.

Odusanya, S. O., Winter, D., Nolte, L., & Shah, S. (2018). The experience of being a qualified female BME clinical psychologist in a National Health Service: An interpretative phenomenological and repertory grid analysis. *Journal of Constructivist Psychology*, 31(3), 273–291.

Office for National Statistics. (2014). Wealth in Great Britain wave 4: 2012 to 2014. Retrieved from www.ons.gov.uk/peoplepopulationandcommunity/personalandhouseholdfinances/incomeandwealth/compendium/wealthingreatbritainwave4/2012to2014.

Office for National Statistics. (2019). Statistics on alcohol, England 2019. Retrieved from https://digital.nhs.uk/data-and-information/publications/statistical/statistics-on-alcohol/2019/part-1

Office for National Statistics. (2020). Suicides in England and Wales: 2020 registrations. Retrieved from www.ons.gov.uk/peoplepopulationandcommunity/birthsdeathsandmarriages/deaths/bulletins/suicidesintheunitedkingdom/2020registrations#:~:text=5%2C224%20suicides%20were%20registered%20in,10.0%20deaths%20per%20100%2C000%20people.

Onifade, E., Davidson, W., & Campbell, C. (2009). Risk assessment: The predictive validity of the youth level of service case management inventory with African Americans and girls. *Journal of Ethnicity in Criminal Justice*, 7(3), 205–221.

Pendry, N. (2012). Race, racism and systemic supervision. *Journal of Family Therapy*, 34(4), 403–418.

Perry, B. L., Neltner, M., & Allen, T. (2013). A paradox of bias: Racial differences in forensic psychiatric diagnosis and determinations of criminal responsibility. *Race and Social Problems*, 5(4), 239–249.

Pettit, B., & Western, B. (2004). Mass imprisonment and the life course: Race and class inequality in US incarceration. *American Sociological Review*, 69(2), 151–169.

Phinney, J. S. (1989). Stages of ethnic identity development in minority group adolescents. *The Journal of Early Adolescence*, 9(1–2), 34–49.

Phinney, J. S. (1993). A three-stage model of ethnic identity development in adolescence. In M. Bernal & G. Knight (eds), *Ethnic identity: Formation and transmission among Hispanics and other minorities* (pp. 61–79). State University of New York Press.

Pinals, D. A., Packer, I. K., Fisher, W., & Roy-Bujnowski, K. (2004). Relationship between race and ethnicity and forensic clinical triage dispositions. *Psychiatric Services*, 55(8), 873–878.

Pleck, J. H. (1981). *The myth of masculinity*. MIT Press.

Pleck, J. H. (1995). The gender role strain paradigm: An update. In R. F. Levant & W. S. Pollack (eds), *A new psychology of men* (pp. 11–32). Basic Books.

Prison Reform Trust. (2016). Bromley briefings prison factfile: Autumn 2016. Retrieved from www.criminaljusticealliance.org/wp-content/uploads/Bromley-Briefings-Autumn-2016.pdf.

Prison Reform Trust. (2018). Bromley briefings prison factfile: Autumn 2018. Retrieved from www.prisonreformtrust.org.uk/Portals/0/Documents/Bromley%20Briefings/Autumn%202018%20Factfile.pdf.

Ratliff, K. A., & Oishi, S. (2013). Gender differences in implicit self esteem following a romantic partner's success or failure. *Journal of Personality and Social Psychology*, 105, 688–702.

Reidy, D. E., Berke, D. S., Gentile, B., & Zeichner, A. (2014). Man enough? Masculine discrepancy stress and intimate partner violence. *Personality and Individual Differences*, 68, 160–164.

Reidy, D. E., Berke, D. S., Gentile, B., & Zeichner, A. (2015). Masculine discrepancy stress, substance use, assault and injury in a survey of US men. *Injury Prevention*, 22, 370–374.

Reisig, M. D., Holtfreter, K., & Morash, M. (2006). Assessing recidivism risk across female pathways to crime. *Justice Quarterly*, 23, 384–405.

Riggs Romaine, C. L., & Kavanaugh, A. (2019). Risks, benefits, and complexities: Reporting race & ethnicity in forensic mental health reports. *International Journal of Forensic Mental Health*, 18(2), 138–152.

Rivas-Drake, D., Seaton, E. K., Markstrom, C., Quintana, S., Syed, M., Lee, R. M., Schwartz, S. J., Umaña-Taylor, A. J., French, S., Yip, T., & Ethnic and Racial Identity in the 21st Century Study Group. (2014). Ethnic and racial identity in adolescence: Implications for psychosocial, academic, and health outcomes. *Child Development*, 85(1), 40–57.

Roberts, S. O., Bareket-Shavit, C., Dollins, F. A., Goldie, P. D., & Mortenson, E. (2020). Racial inequality in psychological research: Trends of the past and recommendations for the future. *Perspectives on Psychological Science*, 15(6), 1295–1309.

Rosenberger, J. S., & Callanan V. J. (2012). The influence of the media on penal attitudes. *Criminal Justice Review*, 36(4), 435–455.

Ross, L. D. (1977). The intuitive psychologist and his shortcomings: Distortions in the attributional process. In L. Berkowitz (ed.), *Advances in experimental social psychology* (pp. 173–220). Academic Press.

Sánchez, F. J., Westerfeld, J. S., Liu, W. M., & Vilain, E. (2010). Masculine gender role conflict and negative feelings about being gay. *Professional Psychology: Research and Practice*, 41, 104–111.

Savage, M., Devine, F., Cunningham, N., Taylor, M., Li, Y., Hjellbrekke, J., Le Roux, B., Friedman, S., & Miles, A. (2013). A new model of social class? Findings from the BBC's Great British class experiment. *Sociology*, 47, 219–250.

Schneider, D. (2004). *The psychology of stereotyping*. Guilford Press.

Shepherd, S. M., Delgado, R. H., Sherwood, J., & Paradies, Y. (2018). The impact of indigenous cultural identity and cultural engagement on violent offending. *BMC Public Health*, 18(1), 1–7.

Shin, R. Q., Smith, L. C., Welch, J. C., & Ezeofor, I. (2016). Is Allison more likely than Lakisha to receive a callback from counseling professionals? A racism audit study. *The Counseling Psychologist*, 44(8), 1187–1211.

Smedley, A., & Smedley, B. D. (2005). Race as biology is fiction, racism as a social problem is real: Anthropological and historical perspectives on the social construction of race. *American Psychologist*, 60(1), 16–26.

Smith, L. (2009). Enhancing training and practice in the context of poverty. *Training and Education in Professional Psychology*, 3(2), 84–93.

Smith, L., Allen, A., & Bowen, R. (2010). Expecting the worst: Exploring the associations between poverty and misbehavior. *Journal of Poverty*, 14, 33–54.

Smith, L., Baranowski, K., Allen, A., & Bowen R. (2013) Poverty, crime seriousness, and the 'politics of disgust'. *Journal of Poverty*, 17(4), 375–393.

Snowden, R. J., Gray, N. S., & Taylor, J. (2010). Risk assessment for future violence in individuals from an ethnic minority group. *International Journal of Forensic Mental Health*, 9(2), 118–123.

Social Exclusion Unit. (2002). Reducing reoffending by ex-prisoners. Retrieved from www.bristol.ac.uk/poverty/downloads/keyofficialdocuments/Reducing%20Reoffending.pdf.

Spence, N. (2012). Cultural competence: Social class – the forgotten component. *Clinical Psychology Forum*, 230, 36–39.

Stephan, W. G., & Stephan, C. W. (2000) An integrated threat theory of prejudice. In S. Oskamp (ed.), *Reducing prejudice and discrimination* (pp. 23–46). Lawrence Erlbaum.

Stephens, N. M., Markus, H. M., & Phillips, L. T. (2014). Social class culture cycles: How three gateway contexts shape selves and fuel inequality. *Annual Review of Psychology*, 65, 611–634.

Sue, D. W., Arredondo, P., & McDavis, R. J. (1992). Multicultural counseling competencies and standards: A call to the profession. *Journal of Counseling & Development*, 70(4), 477–486.

Sue, S., Zane, N., Nagayama Hall, G. C., & Berger, L. K. (2009). The case for cultural competency in psychotherapeutic interventions. *Annual Review of Psychology*, 60, 525–548.

Sutton Trust and Social Mobility Commission. (2019). Elitist Britain 2019: The educational backgrounds of Britain's leading people. Retrieved from www.gov.uk/government/publications/elitist-britain-2019.

Sykes, G. M. (1958). *The society of captives: A study of a maximum security prison*. Princeton University Press.

Tajfel, H. (1981). *Human groups and social categories*. Cambridge University Press.

Tajfel, H., Turner, J. C., Austin, W. G., & Worchel, S. (1979). An integrative theory of intergroup conflict. In W. G. Austin & S. Worchel (eds), *The social psychology of inter-group relations* (pp. 33–47). Brooks/Cole.

Talley, A. E., & Bettencourt, B. (2008). Evaluations and aggression directed at a gay male target: The role of threat and antigay prejudice. *Journal of Applied Social Psychology*, 38, 647–683.

Townsend, M. J., Kyle, T. K., & Stanford, F. C. (2020). Outcomes of COVID-19: disparities in obesity and by ethnicity/race. *International Journal of Obesity*, 44(9), 1807–1809.

Umaña-Taylor, A. J., Quintana, S. M., Lee, R. M., CrossJr, W. E., Rivas-Drake, D., Schwartz, S. J., & Ethnic and Racial Identity in the 21st Century Study Group. (2014). Ethnic and racial identity during adolescence and into young adulthood: An integrated conceptualization. *Child Development*, 85(1), 21–39.

Vandello, J. A., Bosson, J. K., Cohen, D., Burnaford, R. M., & Weaver, J. R. (2008). Precarious manhood. *Journal of Personality and Social Psychology*, 95, 1325–1339.

Van Dieten, M. (2010). *Moving on: A program for at-risk women – Modules 1 and 6 facilitator's guide*. Hazelden.

Van Dieten, M., & MacKenna, P. (2001). *Moving on facilitator's guide*. Orbis Partners.

Viano, E. C. (1992). The news media and crime victims: The right to know versus the right to privacy. In E. C. Viano (ed.), *Critical issues in victimology: International perspectives* (pp. 24–34). Springer.

Viljoen, J. L., Cochrane, D. M., & Jonnson, M. R. (2018). Do risk assessment tools help manage and reduce risk of violence and reoffending? A systematic review. *Law and Human Behavior*, 42(3), 181–214.

Swavola, E., Riley, K., & Subramanian, R. (2016). *Overlooked: Women in jails in an era of reform*. Vera Institute of Justice.

Walajahi, H., Wilson, D. R., & Hull, S. C. (2019). Constructing identities: the implications of DTC ancestry testing for tribal communities. *Genetics in Medicine*, 21(8), 1744–1750.

Walmsley, R. (2006). World female imprisonment list. Retrieved from www.prisonstudies.org/sites/default/files/resources/downloads/world_female_imprisonment_list_third_edition_0.pdf.

Watkins, D. C., Walker, R. L., & Griffith, D. M. (2010). A meta-study of Black male mental health and well-being. *Journal of Black Psychology*, 36, 303–330.

Wilks-Wiffen S. (2011). *Voice of a child*. Howard League for Penal Reform.

Williams, K., Papadopoulou, V., & Booth, N. (2012). *Prisoners' childhood and family backgrounds: Results from the surveying prisoner crime reduction (SPCR) longitudinal cohort study of prisoners*. Ministry of Justice.

Williams, J. L., Aiyer, S. M., Durkee, M. I., & Tolan, P. H. (2014). The protective role of ethnic identity for urban adolescent males facing multiple stressors. *Journal of Youth and Adolescence*, 43(10), 1728–1741.

Winston, A. S. (2020). Why mainstream research will not end scientific racism in psychology. *Theory & Psychology*, 30(3), 425–430.

Wood, N., & Patel, N. (2017). On addressing 'Whiteness' during clinical psychology training. *South African Journal of Psychology*, 47(3), 280–291.

Yasui, M., & Dishion, T. J. (2007). The ethnic context of child and adolescent problem behavior: Implications for child and family interventions. *Clinical Child and Family Psychology Review*, 10(2), 137–179.

Yoo, H. C., & Lee, R. M. (2008). Does ethnic identity buffer or exacerbate the effects of frequent racial discrimination on situational well-being of Asian Americans? *Journal of Counselling Psychology*, 55(1), 63–74.

Index

Please note that page references to Figures will be in **bold**, while references to Tables are in *italics*. Footnotes will be denoted by the letter 'n' and Note number following the page number.

Abu Ghraib 165
accredited behaviour programmes 146–149
ACEs *see* Adverse Childhood Experiences (ACEs)
actor-observer bias 11, 25–26
adversarial affiliation bias 29–31, 32
Adverse Childhood Experiences (ACEs) 59–60, 144, 188, 194
aggression 116–143; acts of 186–187; aggressor motivation typology *129*; Algebra of Aggression 124–125; assessment 122–126; automatic triggering of goals 94–95; Dynamic Assessment and Management of Situational Aggression (DAMSA) 130–139; dynamic management 128–130; example of aggressive incidents 134; extrinsic instigation 125; idiographic 124; interventions 126–130; intrinsic instigation 125; Maurice (case study) 119, 134, **135**, **136–137**, 138; mixed motive aggressor 117; multimodal therapies 128; nomothetic 124; prediction of 92–95; proactive theories 117, *121*, 124; reactive theories 117, *121*, 124, 125; self-reported 94; single mode therapies 127–128; social psychological theories 120; and social psychology 116–117; theories 120, *121*, 122–126; three-factor model 119; typologies 117–119
Aggression Replacement Training (ART) 127
Akers, R. L. 98–99

Albery, I. P. 61
Alderden, M. A. 222
Algebra of Aggression 124–125
American Psychological Association 220, 229
anchoring effect *see* framing and anchoring effects
Andrews, D. A. 92
anti-social behaviours 51–52, 58, 74, 91, 95, 96, 98, 100, 107, 109, 185, 195; attitudes 97; and personality traits 76
applied psychologists 1; contexts working in 1; multiple clients 6–7; role in social psychology 5–7; as 'scientist-practitioners' 3; *see also* forensic practitioners; multidisciplinary teams
arrest 2, 27, 53, 104, 153, 207, 222; re-arrest 224
Assessment and Classification of Function (ACF) 119
assessment of aggression: approaches 124–126; prospective 123; purposes 122–124; retrospective 123
attachment theory 1, 24, 54, 60, 145, 177, 194, 196; impression management 70–72, 85–87; insecure attachments 71
Attention Deficit Hyperactivity Disorder (ADHD) 2
attitude formation 4
attitudes 9; and beliefs 91–115; formation, role in evaluation of the social world 91–92; identification with criminal others 92; implicit and explicit 91, 92; mere exposure effect 12, 91, 98;

offence-supportive 4; and prediction of aggression 92–95; procriminal attitudes (PCA) 92, 96; rejection of convention 92; social learning theory 98–100; *see also* cognitive dissonance
attribution bias 24
attributions: about behaviour of others 24; attribution bias 35; attribution theory 22–26; errors 11, 25
Aureli, N. 186, 189
Australia: Northern Territory 165, 210; Protective Behaviours Programme 216
authoritarian personality 48
availability heuristic 31–32

Bailey, T. 76
Bals, M. 210
Bandura, A. 99
Banse, R. 92
Bargh, J. A. 94
Barkham, M. 179
Baron, R. A. 117
base rates 41
Beck, A. T. 107
behaviourist perspective, coercion 168–169
beliefs: and attitudes 91–115; core beliefs 107
Beltrani, A. 34
Berkowitz, L. 120
Bernstein, D. P. 197
bias 11; actor-observer bias 11; adversarial affiliation 29–31, 32; Antony (case study) 21, 24, 30, 31–32, 33, 34; attribution 24, 35; blind spot 31; cognitive 4; confirmation 33–34, 40; correspondence 25; de-biasing techniques 37; evidence-based bias-reduction techniques 35; harm associated with 37; hindsight 34–35; human 29; implicit 27; negative 39; own biases, knowing 36–37; pro-male 27; pro-White implicit bias 26–27; racial 27; systemic 211–212; thinking about 35; understanding 38; *see also* attributions
bias mitigation 35–36
Black, Asian and Minority Ethnic (BAME) community 207, 230
Black and Minority Ethnic (BME) community 208
blind spot bias 31
'blinding' techniques 41
Bloom, B. 223, 224
Bochetay, N. 62

Boduszek, D. 52, 53
Bonta, J. 92
Book, A. S. 76
Bourbon, W. T. 169, 172, 173
Bowlby, John 71, 196
brain 80
Braithwaite, J. 105
Brannon, R. 218
Broset Violence Checklist (BVC) 123, 125
Bruce, M. 127
Burgess, R. L. 98

CALMER *see* Controlling Anger and Learning to Manage It – Effective Relapse- Prevention Program (CALMER)
Canada 39, 144, 207; Canadian Correctional Service 160; group work 147, 148; prisons 97
care system, England 216
Carey, T. A. 13, 169
Carlsmith, J. M. 102
Carver, C. 93, 94
CAT *see* Cognitive Analytic Therapy (CAT)
CBT *see* Cognitive Behavioural Therapy (CBT)
Chapman, K. R. 27
Chichinadze, K. 119
Child and Adolescent Mental Health Services (CAMHS) 2
child sexual abuse (CSA) 216, 218, 220, 221
Chow, R. M. 186
Circles of Support and Accountability (COSA) 61, 160
Clark, A. C. 224
Classification of Violence Risk (COVR) 123
coercion: behaviourist perspective 168–169; and control 13, 169–171; criminological perspective 167–168; defining 166; distinguishing from force 174–175; forensic practice 179–180; inevitability of separate perspectives 176; minimising 177–178; perceived 166, 175–176; in prisons 166–167; and social influence 165–182; theory 168; *see also* perceived coercion; perception; Perceptual Control Theory (PCT)
cognition 35, 37, 95–96, 102, 107; deviant 52; distorted 92, 95; and emotions

127; hierarchical model 107; implicit 100; inconsistent 101; and offending 78; social 226; surface 107, 120
Cognitive Analytic Therapy (CAT) 7
cognitive behavioural theorists 108
Cognitive Behavioural Therapy (CBT) 7, 8, 10, 95, 96, 159
cognitive bias 4
cognitive dissonance 12, 23, 91, 100, 105; role in promoting change 101–104; and shame 104–107; *see also* dissonance; New Look dissonance model
cognitive neoassociation theory 120, 125
cognitive representations 50–51
collective group identity 58
collective rationalisation 155
collectivist cultures 26
Collins, P. H. 206
collusion 7
Colvin, M. 167, 168
communication, and information gathering 39–40
community psychology 14
community settings 147–148, 230
compassion focused therapy (CFT) 80
confirmation bias 33–34, 40
consent to treatment 75–76
control: and coercion 13, 169–171; countercontrol 165; defining 169; *see also* Perceptual Control Theory (PCT)
Controlling Anger and Learning to Manage It – Effective Relapse-Prevention Program (CALMER) 127
conviction 2, 27, 123, 161n3; reconvictions 224
Cooper, J. 102–103
core beliefs 107
Correctional Services Accreditation Panel (CSAP) 147
correspondence bias 25
correspondence inference 24–25
COSA *see* Circles of Support and Accountability (COSA)
countercontrol 165
court systems and forensic settings 28–29
covariation model 22–23
Covington, S. 223, 224
criminal behaviour *see* offending behaviour
criminal environments 51
criminal identity 50, 54
criminal justice system 227–229

criminal social identity (CSI) 12, 51–56; aetiology 52–53; critique 53–56; exposure to criminal environments and peers 52; integrated psychosocial model of criminal social identity (IPM-CSI) 12, 51, 52, 53; maintaining criminal behaviour 53; personality traits 53; protection of positive self-evaluations 52–53; psychological crises 51, 52
criminogenic needs 7, 10, 148, 159
CSAP *see* Correctional Services Accreditation Panel (CSAP)
CSI *see* criminal social identity (CSI)
Cullen, F. T. 167
cultural competence 38, 211–215
Cutcliffe, J. R. 138
cycle of violence hypothesis 99

Daffern, M. 119, 123, 125, 128, *129*
DAMSA *see* Dynamic Assessment and Management of Situational Aggression (DAMSA)
DASA *see* Dynamic Appraisal of Situational Aggression (DASA)
David, D. 218
Davis, K. E. 24, 102
De Vogel, V. 223
de-biasing techniques 36, **37**, 40; evidence-based 42
DeBono, K. G. 96
deception 74–78; as a 'high-stake' behaviour 75; self-deception 70, 76
decision-making in forensic settings 28–35; adversarial affiliation 29–31; availability heuristic 31–32; blind spot bias 31; confirmation bias 33–34; framing and anchoring effects 32–33; hindsight bias 34–35; representative heuristic 32
defensive behaviours 85
definitions, social learning theory 98, 99
deindividuation 57
depersonalisation 50, 51, 64n2
deviance 58
Dialectical Behaviour Therapy (DBT) 7
differential association, social learning theory 98, 99
differential reinforcement, social learning theory 98–99
Differential Reinforcement Theory (DRT) 52
discrepancy strain 218

discrimination 197, 204; discriminatory behaviour 184, 204, 205, 226; double 60; long-term exposure 144–145; racial 189, 206–207; and social class 227
disculturation 72
Dishion, T. J. 210
disordered identity 54
dissocial environments 51
dissonance: arousal 102, 104; causes 103; cognitive *see* cognitive dissonance; defining 101–102; idiosyncratic 103; motivation 102; Self-standards model 103; vicarious 104; *see also* attitudes
dissonance arousal 102
dissonance motivation 102
distancing 227
Dodge, K. A. 93–94, 120
dogmatic acceptance 3
Don Dale Youth Detention centre 165, 174
Doorley, J. D. 191
Drennan, G. 79
DRT *see* Differential Reinforcement Theory (DRT)
dual-identity, notion of 57
Durrant, R. 3–4
Duwe, G. 224
Dynamic Appraisal of Situational Aggression (DASA) 12
Dynamic Appraisal of Situational Risk (DASA) 123, 125, *126*
Dynamic Assessment and Management of Situational Aggression (DAMSA) 130–139; dynamic management planning *136*; as a dynamic management planning instrument 134–135, **136**; exploratory questions *133*; imminent risk prediction and formulation *132–133*; likely motives for aggressor motivations, eliciting *133*; as a prospective and retrospective risk assessment 131, **132**, **133**, *133*, 134; for staff supervision and reflection 137–139
dysfunction strain 218

early childhood development 84–85
Early Maladaptive Schemas (EMS) 108
early maladaptive schema (EMS) 196, 197
Edmonds, A. E. 96
'elite' professions 227
Emery, L. G. 71

EMS *see* early maladaptive schema (EMS)
entitativity 51
epistemic pluralism 3
Esposito-Smythers, C. 191
essentialism 55
ethnic identity 13, **209**; achievement of 209; and concept of self 209–210; Phinney's stages of ethnic identity development **209**, 211; positive 208–211, 230; unexamined 208–209
ethnicity *see* race and ethnicity; Black, Asian and Minority Ethnic (BAME) community; Black and Minority Ethnic (BME) community
ethnocentrism 48
ethnographic practice 11
Evans, A. M. 214
evidence-based tools 40–41
exemplification 81
explicit attitudes (EAs) 92, 95
expressive independence 225
extremism 58–59

Felitti, V. J. 188
Female Additional Manual (FAM) 223
Festinger, L. 101, 102
field studies 29–30
focused-ethnography 79
force, compared with coercion 175–176
forensic practice/settings: assessment 31, 212; complexity of social contexts 9; cultural competence 38; decision-making in 28–35; diversity of population 208; in forensic settings 72–74; hospitals 100–101; impression management 70, 86–88; inherent social nature of 3; mental health issues 1, 31, 212, 220, 228; multiple clients 9–10; role of social psychology in 7–11; social identity approaches 56–59; theory–practice connections, importance 3–4; *see also* forensic practitioners
forensic practitioners 1, 4, 12, 13, 178, 190, 198; attributions and biases 21, 28, 30, 35, 40, 42; impression management 70, 73, 81, 82, 86, 87; social identity theories 54, 59, 63
forensic rehabilitation 147–148
Foucault, M. 168
foundation theories, aggression 120, *121*
framing and anchoring effects 32–33
Frings, D. 61

frustration-aggression hypothesis 120, 125
functional antagonism 57
fundamental attribution error 222

gangs 9
Gannon, T. A. 152
Gaylin, W. 144
Gebhard, K. T. 219
gender 215–224; defining gender and responsiveness 223; disproportional impact 216; gender role strain 13, 216–218; ideologies 217–218; implications for forensic practice 218–221, 222–224; men and boys 215–221; Moving On initiative 147, 224; stereotyping 9; women and girls 221–224
Gender Role Strain Paradigm (GRSP) 216–218, 220
general aggression model (GAM) 120, 122
'Germans are different' hypothesis 48
Gillard, N. D. 75, 76
Gilligan, J. 85
GLM see Good Lives Model (GLM)
Goffman, E. 78; *The Presentation of Self in Everyday Life* 72, 79
Golding, K. 84–85, 86
Good Lives Model (GLM) 4, 8, 60, 160
Govan, C. L. 186
Gowensmith, W. N. 29
Gozna, L. F. 75
Griffith, R. L. 40
group formation 12, 149–154; adjourning 150; forming phase 149, 152; investigation phase 150; maintenance phase 151; norming 149–150; performing 150; socialisation phase 150–151; storming 149, 151–152, 153
groups 9, 12, 144–164; Amanda (case study) 157, 158; collective group 'personality' 57; criticism 78; delivery of accredited behaviour programmes 146–149; effects 56; formation *see* group formation; groupwork during detention 146–154; interventions by 79; membership 47, 49, 53, 56, 58; norms 58; polarisation 12, 158–159, 160; praise 78; professional group decision-making and exit from the system 154–159; prototypes 54; Rashid (case study) 145, 146; remembrance 151; re-socialisation 151; small and large group work (circles) 79; socialisation 12; Steve (case study) 153; *see also* in-groups; out-groups
groupthink 12, 154–158; social psychological theories 160; symptoms 156
GRSP *see* Gender Role Strain Paradigm (GRSP)
Guantánamo Bay 165
guilt, and shame 57, 70, 84, 86
Gypsy, Roma and Travellers of Irish Heritage 208

hard interdependence 225
Hart, C. M. 150
Hartigan, S. E. 76
Harvey, J. 10, 79, 191
Haslam, A. 10
HCR-20 (Historical-Clinical-Risk 20) 30, 34, 125, 223; HCR-20v3 124; impression management 75, 76
Heider, F. 22
Hemmings, C. 214
heuristics 11, 24; availability 31–32; representativeness 32
Higgins, E. T. 82, 84, 85, 92–93
hindsight bias 34–35
HM Prison and Probation Service (HMPPS) 161n5
Hoge, R. D. 101
Hogg, M. 51, 59
Holden Psychological Screening Inventory 76
Howells, K. 128
Hubbard, D. J. 104
Hughes, D. 84–85, 86
hypothesis testing 39, 40

IAT *see* Implicit Association Test (IAT)
identity: collective 58, 63; criminal 50, 54; disordered 54; dual-identity, notion of 57; ethnic 13, 208; personal 49, 58; pro-social 60, 61; uncertainty-identity theory 59; *see also* Integrated Psychosocial Model of Criminal Social Identity (IPM-CSI); social identity approaches
identity fusion 58
identity performance 54
imitation, social learning theory 99
Immigration Act (2014) 168
Implicit Association Test (IAT) 27, 36, 96, 109, 214
implicit attitudes (IA) 92, 93, 95
implicit bias 27
implicit tests 109

246 Index

impression construction 81–82
impression management 12, 69–90; adaptive and mal-adaptive functions 85; adjustment approach 77; attachment theory 70–72, 85–87; and deception 74–78; defensiveness approach 77; defining 72; forensic practitioners 86–88; forming impressions 26–28; and personality traits 76; positive, on risk assessments 76; public and private self 78–81; Ryan (case study) 69–70, 73; scales 77; self-discrepancy and experiences of shame and guilt 82, 83–84, 85–86; self-guides 83, 84–85; self-states 83, 84–85; strategic impression management construction, motivations for 81–82; strategies 75, 82; *see also* self-presentation
impression motivation 81
imprisonment: non-violent crimes 221, 222; and ostracism 187–190; parental 190; as 'social death' 183, 190; US incarceration rates 221; of women 221; *see also* prisons
Improving Access to Psychological Therapies (IAPT) 215
incarceration, length of 54
indeterminate public protection (IPP) 153
individualistic cultures 26
induction training 38
information gathering 39–40
informational social influence 58
ingratiation 81
in-groups: affect 52, 53; in-group favouritism 48, 49, 50; identification with 50; members 49; and prejudice 205; prototypical members 51; ties 52
Inhabitations against Aggression 125
Instigation to Aggression (A) 125
institutional distancing 227
Integrated Psychosocial Model of Criminal Social Identity (IPM-CSI) 12, 51, 52; critique 53–56
integrated theories, aggression 120, 122
intergroup behaviour, Social Identity Theory 48–50
inter-rater reliability 29
intersectionality 13, 14, 206, 216, 220, 222, 229–230
interventions 4, 7, 8, 11, 12, 15, 28, 36, 96–98, 109, 110, 120, 122, 129, 147–149, 152, 161n3, 172, 179, 220, 222, 223, 229; aggression 126–130; CBT 10, 159; collaborative 139; DASA-informed 130; environmental 130; intimate partner violence (IPV~) 91; legally mandated 166, 178; multimodal 128; parenting 168; pharmacological 130; preventative 134; psychologically informed 101, 183; psychosocial 130, 131, 137; punitive 138; social psychological 139; therapeutic 100, 105, 223; violence-related 91; voluntary 178
intimidation 81
intrinsic instigation 125
introspection illusion 36
invulnerability, illusions of 155
IPM-CSI *see* integrated psychosocial model of criminal social identity (IPM-CSI)
Ireland, T. O. 99–100
Iwamasa, G. Y. 214

Janis, I. L. 154, 155
Johnson, A. J. 230
Jones, E. E. 23, 24, 69, 81, 102
judgement, personal standards 103

Kaunomäki, J. 130
Kelley, H. H. 22–23
Kennedy, J. F. 155
Knapton, H. 187
Kowalski, R. H. 74, 81–82
Kraus, M. W. 226
Kugler, M. B. 36
Kyprianides, A. 189

Lamm, H. 158–159
Lammy Review (2017) 207
Leary, M. R. 72, 74, 81–82, 187
LeBourgeois, H. I. 34
Leuschner, V. 187
Levant, R. F. 220
Level of Supervision Inventory-Revised (LSI-R) 223
Levine, J. M. 150
Liang, C. T. H. 220
Liebling, A. 191
Linder, D. E. 102
linear sequential unmasking 41

MAPPA *see* multi-agency public protection arrangements (MAPPA)
Marken, R. M. 169, 170–171
Marquart, J. W. 166–167
Marshall, W. L. 148–149

Martin, K. 39, 40
Masculine Gender Role Discrepancy Scale 218
masculinity 215, 217, 219, 220; ideal 218; notions of 220; socialisation 221; threats to 218, 219
Masculinity and Shame Questionnaire 219
Maycock, M. 230
Mead, G. H. 72
Megargee, E. I. 124–125
Mental Capacity Act (2005) 175
Mental Health Act (1987) 6, 155, 161n4
mental health issues 54, 59, 60, 64, 145, 190, 192, 195, 222, 224; forensic assessments 31, 212; forensic settings 116, 139, 176; services/support 1, 8, 214, 220, 228
Mental Health Review Tribunal 30
mental health tribunal (MHT) 154, 161n4
mere exposure effect 12, 91, 98
meta-contract principle 58
meta-theory 47, 48, 54
Milgram, S. 5
mitigation: bias mitigation 35–36; in forensic practice 36–42
Monin, B. 104
monitoring logs 38–39
Moreland, R. L. 150
Moving On initiative 147, 224
MST *see* Multi-Systemic Therapy (MST)
Mullen, R. 174
multi-agency public protection arrangements (MAPPA) 159, 161n5
multidisciplinary teams 6, 8, 24, **25**, 26
multifactorial classification systems 119
multiple assessors 41–42
multi-systemic therapy (MST) 7, 128
Munoz, R. T. 194
Murrie, D. 30
Myers, D. G. 158–159

naïve (common-sense) psychology 23
narcissism 77
National Health Service (NHS) 5
National Institute for Health and Care Excellence 117
natural disasters 102
Neal, T. 41
needs principle 8, 148
Neuman, G. A. 150
neutralisation techniques 92

New Look dissonance model 102–103
Newton-Howes, G. 174
NHS Trust 8
Nisbett, R. E. 23
Nolan, K. A. 119
normative consequences of behaviour 103
Norton, M. I. 104
Novaco, R. W. 127
Nussbaum, D. 119

Occupational Therapist (OT) 158
offence-focused work, versus support for mental health disorders 8
Offender Manager (OM) 153, 154
Offender Personality Disorder (OPD) 100
offending behaviour: development of 98, 99–100, 101; intergenerational transmission of 54; non-violent 221, 222; public and private self 78; sexual offences 57–58, 75; treatment programmes 95–97
Ogloff, J. R. 123, 125, 128
OM *see* Offender Manager (OM)
operant conditioning model 99
operant theory principles 98
organisational citizenship behaviours (OCBs) 63
ostracism 9, 183–203; Adam (case study) 191–197; application of model 194–195; and belonging 183–184; clinical implications 191–197; concept 183, 191, 197; and imprisonment 187–190; intervention 196–197; multi-layered 187, **188**; ostracism-based experiences 184; reflective stage 195; reflexive stage 194–195; resignation stage 195; social death 183, 190; suicidal ideation 190–191; temporal need-threat model 185–186; vignette 191–197; and violence 186–187
out-groups 49, 50, 51, 55, 62; social alienation 60, 61
Owen, B. 224

Parole Board of England and Wales 22, 157
Paulhus, D. L. 70
Paulhus Deception Scale (PDS) 76
PCT *see* Perceptual Control Theory (PCT)
PDS *see* Paulhus Deception Scale (PDS)
Pendry, N. 214

perceived coercion 166, 174, 177, 178, 179; origins 175–176; *see also* coercion; Perceptual Control Theory (PCT)
perception 93, 94, 169; counter disturbances to perceptual states 177; and 'difference' 62, 63; of forensic populations 55–56; foresight and hindsight 34–35; impression management 82; of intent 117; negative, of outgroups 62, 63; others' perceptions of us 78; perceptual states 177; public and political 4; sensory 107; social 94; and social identity 47, 49, 52, 53
Perceptual Control Theory (PCT) 13, 169, 170; and basic feedback loop **170**, 173; considering social interaction from perspective of 171–173; functional or generative models 173; general implications for implementing services 178–179; implications of models for understanding social interactions 173–174; mandated versus voluntary treatment 178–179; perspective improving forensic practice 177–178; as Test for the Controlled Variable 177, 178
performance 73, 196; academic 227; and class 225; defining 72; identity 54; self-presentation 70, 72, 74, 81
personal consequences of behaviour 103
personal identity 49, 58
personal responsibility 102
personality: application of schemas to personality theory 107–109; maladaptive dispositional traits 25; traits 25, 51
Pfundmair, M. 187
Phinney, J. S. 208
Piaget, J. 107
Pittman, T. S. 69, 81
Pleck, J. H. 216, 217, 218
polarisation 12, 158–159, 160
poverty 9
Power Threat Meaning 8
Powers, W. T. 171
prejudice 13, 28, 48, 96, 145, 206; against Black communities 27; characteristics 204–205; defining 204; and discrimination 212–215; expression of 205; extreme 92; and in-groups 205; and stereotyping 204, 210, 216, 217, 219–222, 224, 225, 227, 229, 230; *see also* groups; in-groups; out-groups; race and ethnicity; stereotyping
priming research 93
Prison Fellowship 79
prisons 6, 116, 152, 176, 188, 222; Aboriginal and Torres Strait Islander prisoners 210; in Canada 97; coercion in 166–167; guards 156; imprisonment as 'social death' 183, 190; 'mainstream' population 75; prison experiments 5–6, 156; prison-based therapeutic communities 97; and punitive ostracism 188; and self-presentation 70, 73, 80, 85; in the UK 8, 147, 160, 166; *see also* community settings; secure settings
Probation Service in England and Wales 147
procriminal attitudes (PCA) 92, 96
Pronin, E. 36
prototypes 50–51, 59, 62; criminal group 53, 54
psychological crises 51, 52, 54, 60
Psychology Inventory of Criminal Thinking Styles (PICTS) 76
psychopathy 30, 75, 76
Psychopathy Check List–Revised (PCL-R) 30, 76
psychosocial intervention strategies *129*
public and private self 12, 78–81
Public Health model 14
Pupil Referral Unit (PRU) 2
Puzzle Test 109

Quanbeck, C. D. 119

race and ethnicity 206–215; becoming a culturally competent clinician 211–215; custodial sentences 222; discrimination 189; in forensic settings 207–208; racial and ethnic awareness in clinical practice 212, **213**, 214–215; racism 9; schizophrenia, overdiagnosis among Black communities 26; systemic bias 211–212; therapy dialogue **213**; *see also* ethnic identity; racialised minorities
Rachlinski, J. 27
racialised minorities 207; Black African 26; Black British 26; Black African American 26, implicit biases against 27, 28; overrepresentation in justice system 207; women 222; *see also* Black, Asian and Minority Ethnic (BAME) community; Black and

Minority Ethnic (BME) community; ethnicity
radicalisation 57
Ramesh, T. 123
Realistic Conflict Theory 48
Reasoning and Rehabilitation Programme (R&R) 147
recidivism 109
reflective practitioners 3
Reicher, S. D. 10
Reidy, D. E. 218, 220
reinforcement trap 168
reintegrative shaming 8, 105, **106**
Reisig, M. D. 223
rejection-based experiences 184
representativeness heuristic 32
responsivity principle 148
restorative justice (RJ) 8, 104
Riahi, S. 138
Richardson, D. R. 117
risk assessment 9, 10, 70, 76, 116, 123, 124; attributions and biases 21, 28, 33, 34, 38, 41; and gender 224; tools 29–30, 212, 223
risk minimisation 76
risk principle 148
risk state 123
risk status 123
Risk-Need-Responsivity (RNR) model 7, 92, 148, 149, 159; limitations of 8–9
RNR *see* Risk-Need-Responsivity (RNR) model
Roberts, S. 212
Rogers, R. 75, 76
role-stripping 72
Rosenberger, V. 12
Rudman, L. A. 96
Ruscio, J. 32
Rwanda, genocide in 206

Sámi community, northern Scandinavia 210
Scheithauer, H. 187
Schema Therapy (ST) 7, 196
schemas: application to personality theory 107–109; concept 107; domains 108; formation 107; maladaptive 108
schizophrenia, overdiagnosis among Black communities 26
SCT *see* Self-Categorisation Theory (SCT)
secure settings 6, 10, 54, 70, 73, 80, 82, 87, 183; forensic 72; low, medium and high 8; *see also* prisons

self: actual and ideal 12, 52, 70; 'authentic' presentations of 80; as a collaborative product 72; core 79; defined as a social construct 48; looking glass 78; performance of 70; public and private 12, 70, 78–81; sense of 103, 226; vulnerable 79
self-categorisation theory (SCT) 11, 48, 50–51, 62, 64n2
self-concept 78, 82
self-disclosure 87
self-discrepancy theory 12, 70, 82, 83–84, 85–86; actual and ideal self 12, 52
self-esteem 33, 49–53, 101, 103, 104, 108, 185, 187, 210; assessment 223; enhancing 49, 50, 52; high 103; low 9, 52–53, 145, 148, 149; positive 49, 56, 186; threats to 57
self-monitoring 110n1
self-presentation 12, 19, 70–74, 78–82, 85; adaptive 73, 80; as deceptive 74; defining in forensic settings 72–74; differential adaptive 73; forms 80, 81; mask of 74, 81; performance 70, 72, 74, 81; strategic 81; theatre analogy 72; *see also* impression management
self-promotion 81
self-reflexivity 214
self-serving attribution bias 11
self-standard model 103
self-states 83, 84–85
sentences 9, 27, 28, 82, 152, 161n3; additional 34, 153, 154; commencing 145; community 73, 106; custodial 101, 106, 146, 165, 222; female prisoners 221; harsh 27, 58, 193, 222; indeterminate 2, 21, 154; life 145, 146; long 192; mandated 178; original 154; planning 8, 78, 147, 153, 174, 175, 176, 177, 178; recalling 94; short 145, 153, 219; young people 101, 145
Settles, I. H. 14
Sex Offender group initiatives 147
Sex Offender Treatment Programmes 75, 160
sexual offences 54, 75, 160, 224; assault allegations 222; child sexual abuse (CSA) 216, 218, 220, 221; rape 145
shame 8, 74, 75, 77, 145, 186; and cognitive dissonance 104–107; deep-rooted 219; and embarrassment 83; experience of 12, 80, 84, 86; and guilt

57, 70, 84, 86; internalised 219, 230; re-integrative 106; shield against 85; unprocessed 86
shaming: reintegrative 8, 105, **106**; stigmatic 105
Sherif, M. 48
SIP *see* Social Information Processing (SIP) theory
SIT (Social Identity Theory) *see* social identity approaches
Skinner, B. F. 99; *Science and Human Behavior* 169
slavery 165
Slovic, P. 39
Smedley, K. 10–11
Smith, C. A. 99–100
Smith, L. 229
social alienation 60, 61
social approval 77
social class 224–229; classism 228; and criminal justice system 227–229; social psychology of 225–227
social contexts 9, 10
social disapproval 77
social exclusion 183–184
social identity approaches 9, 11, 47–68, 104, 208; application to forensic populations 56–59; criminal social identity 12, 55; early trauma and social alienation as foci for 'outgroup' status 59–61; engaging with online groups and extreme behaviour 56–59; ethical dilemmas associated with 55–56; extremism 58–59; multiple social identities 48–49; and Nazi Germany 47–48; organisational implications for forensic services 61–63; Social Identity Theory (SIT) 48–51, 54–55, 56, 63; staff conflict, performance and retention 62–63; team composition 62
Social Identity Model of Deindividuation Effects (SIDE) 58
social influence 9
Social Information Processing (SIP) theory 120
social justice 15
social learning theory 12, 91, 98–100, 120, 125
social ostracism *see* ostracism
social psychology 15; and aggression 116–117; applied psychologist's role in 5–7; groupthink 160; role in forensic practice 7–11; theory *see* social psychology theory

social psychology theory: contributing to sense-making 22–28; theory–practice connections 3–5
socialisation 149, 220, 221
socio-cognitive 'deficits' 95
socioeconomic status (SES) 224, 225, 227
Sommer, F. 187
Space Shuttle Challenger disaster 155
SPE *see* Stanford Prison Experiment (SPE)
ST *see* Schema Therapy (ST)
Stanford Prison Experiment (SPE) 5–6, 156
Stephens, N. M. 225
stereotyping 13, 26, 204–240; class 224–229; gender 9, 215–224; in-groups and out-groups 155; and prejudice 204, 210, 216, 217, 219–222, 224, 225, 227, 229, 230; and prototypes 51; race and ethnicity 206–215; social stereotypes 205; unconscious 27; *see also* groups; in-groups; out-groups
Stewart, R. L. 96
stimulus response (S-R) bond 168
Stone, J. 102–103
structured professional judgement (SPJ) 75
Sue, S. 211
suicidal ideation 190–191
suicide, male 215
supplication 81
Sutherland, E. H. 98–99
Sycamore Tree Programme 79
Sykes, G. M> 230
symbolic interactionism 72

Tait, S. 156
Tangney, J. P. 84
Taylor, K. 130
temporal need-threat model of ostracism: reflective stage 185; reflexive stage 185; resignation stage 186
Test for the Controlled Variable (TCV) 177, 178
theoretical illiteracy 3
theory–practice connections, importance 3–5
Therapeutic Communities 128
therapeutic milieu 96, 97, 101
Thinking for a Change (TFAC) 147
Tice, D. M. 78
Tiedens, L. Z. 186
total institutions 72
Townsend, M. J. 210
training and supervision 38

trauma 9, 12, 14, 28, 144, 188, 194, 208, 210, 211; attachment 24; complex 158; early/childhood trauma 59–61, 108, 153; histories 61, 100, 148, 159; of parental imprisonment 190; race-based traumatisation 214; role in clients' histories 1; trauma strain 218, 220; trauma-informed care (TIC) 60, 224, 230; trauma-informed practice 8, 10, 42, 100, 159; violence 224
Tuckman, B. W. 149–150
Tully, R. 76
Turnbull, M. 165

Ullman, S. E. 222
uncertainty-identity theory 59
universities, high-status 226, 227
Uziel, L. 77

Van de Weijer, S. G. 190
Van Vugt, M. 150
Vander Ven, T. 167
vicarious dissonance 104
violence, and ostracism 186–187
violence reduction 4, 78; Violence Reduction Programme (VRP) 97, 147
Violence Risk Screening-10 (V-RISK-10) 123
Violent Offenders Treatment Programme (VOTP) 128
VOTP *see* Violent Offenders Treatment Programme (VOTP)

Ward, T. 1, 3–4
Weiner, B. 23
Welsh, B. C. 101
'What Works?' debate 8, 147, 148
white fragility concept 210
Williams, C. A. 191
Williams, M. M. 76
Wilson, T. D. 101
women and girls 221–224; gender-responsive services 224
Word Completion Task 109
Wright, J. 150

Yasui, M. 210
Young, J. E. 107, 196, 197
Young Offender Institutions (YOIs) 101, 145, 192
Youth Offending Team (YOT) 2, 73

Zajonc, R. B. 96, 98
Zapf, P. A. 31, 36
Zelli, A. 94
Zimbardo, P. G. xiii, 5–6, 11, 156

Taylor & Francis eBooks

www.taylorfrancis.com

A single destination for eBooks from Taylor & Francis with increased functionality and an improved user experience to meet the needs of our customers.

90,000+ eBooks of award-winning academic content in Humanities, Social Science, Science, Technology, Engineering, and Medical written by a global network of editors and authors.

TAYLOR & FRANCIS EBOOKS OFFERS:

- A streamlined experience for our library customers
- A single point of discovery for all of our eBook content
- Improved search and discovery of content at both book and chapter level

REQUEST A FREE TRIAL
support@taylorfrancis.com

Routledge — Taylor & Francis Group
CRC Press — Taylor & Francis Group

Lightning Source UK Ltd.
Milton Keynes UK
UKHW030613301222
414415UK00015B/51